This seminal volume defines a new theoretical terroir by addressing the complex issue of geographical indicators on a global scale, tackling the multiple ways in which GIs have been used (and misused) around the world.

—**Lawrence Busch**, University Distinguished Professor Emeritus, Michigan State University, USA

This volume engages critically with ideas, grounded practices, and effects of geographic indicators applied to food. In problematizing the concepts of tradition, locale, and market relations, this volume advances a timely critique of the emancipatory power of a broad range of agri-food alternatives.

—**Steven A. Wolf**, President of the International Sociological Association Research Committee on Sociology of agriculture and Food (RC-40) and Associate Professor at Cornell University, USA

This excellent book provides readers with an insightful analysis of the critical and too often neglected issue of the actual impact of Geographical Indication. I highly recommend it to readers who wish to understand the desirable and undesirable consequences engendered by the implementation of Geographical Indication.

—**Andrea Marescotti**, *PhD*, Associate Professor of Agri-Food Economics, Department of Economics and Management University of Florence, Italy

This book takes a post-industrial view of Geographical Indications in global markets, providing empirical basis for readers to question whether various country contexts allow this policy framework to achieve hoped-for ideals, namely rural development, heritage protections, and protection of biodiversity.

—**Dominique Barjolle**, Senior Researcher and Lecturer, Swiss Federal Institute of Technology (ETHZ), Agroecosystem Group

Geographical Indication and Global Agri-food

This book addresses the relevance of geographical indication (GI) as a tool for local and socioeconomic development and democratization of agri-food, with case studies from Asia, Europe and the Americas.

A GI is a sign used on products that have a specific geographical origin and possess a reputation or qualities that are due to that origin. It provides a way for businesses to not only leverage the value of their geographically unique products but also inform and attract consumers. A highly contested topic, GI is praised as a tool for the revitalization of agricultural communities while also criticized for being an instrument exploited by global corporate forces to promote their interests. There are concerns that the promotion of GI may hamper the establishment of democratic forms of development. The contributing authors address this topic by offering theoretically informed investigations of GI from around the world. The book includes case studies that analyze green tea in Japan, olive oil in Turkey, dried fish in Norway, French wine and Mexican Mezcal. It also places GI in the broader context of the evolution and trends of agri-food under neoliberal globalization.

The book will be of interest to researchers, policymakers and students in agri-food studies, sociology of food and agriculture, geography, agricultural and rural economics, environmental and intellectual property law and social development.

Alessandro Bonanno is Texas State University System Regents' professor and distinguished professor of sociology at Sam Houston State University, US.

Kae Sekine is associate professor of economics at Aichi Gakuin University, Japan.

Hart N. Feuer is junior associate professor of rural sociology at Kyoto University, Japan.

Other books in the Earthscan Food and Agriculture Series

Organic Food and Farming in China
Top-down and Bottom-up Ecological Initiatives
Steffanie Scott, Zhenzhong Si, Theresa Schumilas and Aijuan Chen

Farming, Food and Nature
A Sustainable Future for Animals, People and the Environment
Edited by Joyce D'Silva and Carol McKenna

Governing Sustainable Seafood
Peter Oosterveer and Simon Bush

Farming Systems and Food Security in Africa
Priorities for Science and Policy Under Global Change
Edited by John Dixon, Dennis P. Garrity, Jean-Marc Boffa, Timothy Olalekan Williams, Tilahun Amede with Christopher Auricht, Rosemary Lott and George Mburathi

Consumers, Meat and Animal Products
Policies, Regulations and Marketing
Terence J. Centner

Gender, Agriculture and Agrarian Transformations
Changing Relations in Africa, Latin America and Asia
Edited by Carolyn E. Sachs

A Global Corporate Trust for Agroecological Integrity
New Agriculture in a World of Legitimate Eco-states
John W. Head

Geographical Indication and Global Agri-Food
Development and Democratization
Edited by Alessandro Bonanno, Kae Sekine and Hart N. Feuer

For further details, please visit the series page on the Routledge website: www.routledge.com/books/series/ECEFA/

Geographical Indication and Global Agri-Food

Development and Democratization

Edited by Alessandro Bonanno, Kae Sekine and Hart N. Feuer

Routledge
Taylor & Francis Group
LONDON AND NEW YORK

First published 2020 by Routledge

2 Park Square, Milton Park, Abingdon, Oxon, OX14 4RN
605 Third Avenue, New York, NY 10017

Routledge is an imprint of the Taylor & Francis Group, an informa business

First issued in paperback 2020

British Library Cataloguing-in-Publication Data
A catalogue record for this book is available from the British Library

Library of Congress Cataloging-in-Publication Data
A catalog record for this book has been requested

ISBN: 978-1-138-60047-8 (hbk)
ISBN: 978-0-367-72711-6 (pbk)

Typeset in Bembo
by Apex CoVantage, LLC

Contents

Acknowledgments

This volume originated from our previous work on alternative forms of food production and consumption and research on geographical indication. Specifically, this book is based on the sessions that we organized at the XIX World Congress of Sociology held in Toronto, Canada, in July 2018 under the sponsorship of the Research Committee on Sociology of Agriculture and Food (RC-40). We would like to thank all the members of RC-40 that made these original sessions possible and those who participated in the discussions at these sessions. We would also like to acknowledge the support and comments about GI that Dr. Florence Tartanac and her team at the Nutrition and Food System Division of the United Nations (UN) Food and Agriculture Organization (FAO) provided to Kae Sekine during her 2018–2019 sabbatical stay at FAO. Our gratitude goes also to the Lotte Foundation of Japan for their financial support enabling us to carry out book preparations in person and refine the book contents at a number of international events.

Alessandro Bonanno
Kae Sekine
Hart N. Feuer

Contributors

Virginie Amilien is research professor in the Department of Consumption Research (SIFO) at the Oslo Metropolitan University (Norway) that she joined in 1999. She works as a cultural historian and ethnographer on issues related to food, consumption and culture, with a special emphasis on Norwegian food culture, national identity, food system in tourism, food practices, food consumption, culture of consumption, food and migration and especially local and terroir food products. Together with Erling Krogh, she edited the book *The Cultivated Food* (2007), which is the first anthology about food culture published in Norway. She has been heavily involved in the international scientific journal *Anthropology of Food* (https://aof.revues.org/).

Donna Appavoo (PhD) is an instructor in the Department of Sociology and Chang School at Ryerson University in Toronto, Canada. She completed her doctoral studies in human geography at the University of Waterloo in Canada. Dr. Appavoo has a background in nutrition and public health. The focus of her academic work is Indigenous foodways, food and health environments in rural and remote settings in Canada. Dr. Appavoo has published several articles on the nutritional composition of Indigenous foods that have since been considered as foundational for nutrition studies among Indigenous populations in Canada.

Romain Blancaneaux holds a PhD in political science and is associate researcher at the Centre Émile Durkheim (CED), Sciences Po Bordeaux, France. Looking specifically at wine, his research focuses on the political regulation of the economy. He has published articles and book chapters on the impact of European Union (EU) policy instruments on the wine sector, on sectorial economic and collective strategies, and on the relationship between national and European Parliaments opposing the European Commission's reform of the EU wine policy.

Alessandro Bonanno (PhD sociology, University of Kentucky, 1985) is Texas State University System Regents' professor and distinguished professor of sociology at Sam Houston State University, US. Focusing on agri-food, he studies the neoliberal globalization of the economy and society.

In particular, he investigates the impact that neoliberal globalization has on democracy, labor relations and the emancipatory options of subordinate groups. Dr. Bonanno is the author of numerous publications that appeared in English and other major languages. His most recent books are *The legitimation Crisis of Neoliberalism* (2017), *Resistance to the Neoliberal Agri-Food Regime: A Critical Analysis* (2018) and *State Capitalism under Neoliberalism: The Case of Agriculture and Food in Brazil* (2019).

David Rodolfo Domínguez Arista is a PhD candidate in sociology in the Department of Rural Sociology at the University of Chapingo, Mexico. His research focuses on localized agri-food systems and local markets, particularly the denomination of origin of mezcal and the effects it has had on peasant and Indigenous communities in Southern Mexico. He has published "The Avatars of Artisanal Production of Mezcal: the Denomination of Origin, Exclusion and Alternatives" (2016), "Organic Agriculture in Michoacán and Organic Avocado. Ecological Certification as a Means of Export to the International Market" (2018).

Angga Dwiartama (PhD) is assistant professor at the School of Life Sciences and Technology at the Institut Teknologi Bandung (ITB), Indonesia. Dr. Dwiartama's works focuses on the geography of agri-food, probing issues such as the global value chains/global production networks, agri-environmental governance, local food movements and food security discourses. His research has exclusively been conducted in Indonesia, covering commodity analyses of rice, cocoa, coffee and sugar.

Hart N. Feuer is a junior associate professor at the Graduate School of Agriculture, Kyoto University, Japan. He has a PhD in agricultural sociology from the University of Bonn. His work centers on the transition of agri-food systems in Asia, with a focus on nutrition, culinary heritage, lifelong food skills, food trade and rural-urban food policies. Countries of focus include Cambodia, Japan, Malaysia and Vietnam. Contact: feuer.hartnadav.4e@kyoto-u.ac.jp

Atle Wehn Hegnes is a researcher at NIBIO – the Norwegian Institute of Bioeconomy Research. He holds a PhD in sociology from the University of Oslo and has worked at the National Institute for Consumer Research in Norway. He has experience with qualitative and interdisciplinary research on local food, food culture, agriculture, bioenergy, wood waste, consumption, policies and regulations. In particular, he focuses on the interlinkages of practices and their transformative potential to sociocultural change within the framework of bioeconomy.

Monika Korzun is currently completing her PhD in the rural studies program at the University in Guelph in Canada. Korzun's research interests focus largely on examining the food system from a farmer's perspective. Her dissertation focuses on the experiences of farmers partaking in value-based food supply chains in Southern Ontario. Korzun has published on a variety of topics, including multifunctionality, trading agreements and the alternative food system. She has been teaching food-related courses since 2012 and wishes to inspire others to improve the world through food.

Gilberto Mascarenhas is researcher at the Center "Market, Networks and Values" of the CPDA of the Federal University of Rio de Janeiro, Brazil. He focuses on markets for family farming, territorial development and social networks. Among his current areas of interest are the social construction of markets, sustainable gastronomy, geographical indications, alternative food networks and sociotechnical networks. Dr. Mascarenhas is one of the coordinators of the Brazilian Network of Localized Agrifood Systems (Sial Brasil). His most recent publications include a book, written in partnership with Paulo Niederle and John Wilkinson called *O Sabor da Origem* (2016), and articles *The Sustainable Gastronomy of Paraty* (2018), and *The Social Construction of Quality in Agri-Food Localized Systems: the case of Montpeyroux, France* (2018).

Paulo Niederle is assistant professor of rural and economic sociology at the Federal University of Rio Grande do Sul, Porto Alegre, Brazil. His research focuses on food markets, certification systems, organic agriculture, public policies and family farming. Dr. Niederle is vice-president of the Latin American Rural Sociology Association, Latin American representative in the Research Committee on Sociology of Agriculture and Food of the International Sociological Association and member of the FAO-CLACSO Working Group on Food Security. His most recent book, written in partnership with Valdemar Wesz Jr, is *As novas ordens alimentares* (UFRGS, 2018). In 2016, he also coedited *O Sabor da Origem* with John Wilkinson and Gilberto Mascarenhas.

Derya Nizam is assistant professor of sociology at Izmir University of Economics, Turkey. She received a bachelor's degree and master's degree in sociology from Bogazici University, Turkey, and a PhD in sociology and social policy from the University of Sydney, Australia. Focusing on agricultural and rural transformation, she studies geographical indication systems as an alternative rural development paradigm, together with the various theories of agro-exceptional methodologies in the context of the classic debate on the agrarian question. She received the Ronald Wimberley Best Graduate Student Paper Award by the American Rural Sociological Society in 2012.

Cinzia Piatti (PhD) is lecturer and postdoctoral research associate at the Institute of Social Sciences in Agriculture, under the Societal Transition and Agriculture at the University of Hohenheim, Germany. She focuses her works on the social embeddedness of local food and the value of landscape, specializing on wine and terroir. Her work on the governance of geographical indications appeared in the *British Food Journal* (2017).

Marie-Christine Renard is professor-researcher in the Department of Rural Sociology at the University of Chapingo in Mexico and member of the

Mexican National Researchers System. She has carried out research on peasants and Indian movements in Chiapas, coffee commodity market and quality certification in agri-food. She is currently working on fair trade and other alternative agri-food networks and on localized agri-food systems. She has published numerous articles, book chapters and books on these topics, which have appeared in English, Spanish and French.

Kae Sekine is associate professor in the Graduate School of Economics at Aichi Gakuin University, Nagoya, Japan. Dr. Sekine has researched food standards including geographical indication in Europe and Asia. She was visiting scientist at the French National Institute for Agricultural Research (INRA), Montpellier, during 2007–2010 and at the United Nations Food and Agriculture Organization (FAO), Rome, during 2018–2019. She is the coauthor of *Contradictions of Neoliberal Agri-food: Corporations, Resistance, and Disasters in Japan* (2016).

Anastasiya Shtaltovna is visiting professor at the Graduate School of Agriculture, Kyoto University and a researcher at Centre for International Studies (CÉRIUM), University of Montreal, Canada. For the past decade, she has been working in the field of sustainable development, conducting research in Central Asia, Eastern and Western Europe, and sub-Saharan Africa. She holds a PhD in development studies from the University of Bonn, Germany. Her work and research interests lie in, but are not limited to, rural development, sustainable development, governance, extension services, formal and informal institutions, knowledge and innovation, ethnographic research and comparative studies. Contact: ashtaltovna@yahoo.com

Anelyse M. Weiler is a PhD candidate in the Department of Sociology at the University of Toronto. She is also a 2015 Pierre Elliott Trudeau Foundation Scholar and a Culinaria Research Centre fellow. Her teaching and research explore the convergence of social inequalities and environmental crises across the food chain, with a particular focus on migrant farm workers. She actively contributes to several organizations advocating for food security, migrant justice and decent work. You can read her academic writing and public scholarship at www.anelyseweiler.com

John Wilkinson is associate professor at the Graduate Center: Development, Agriculture and Society (CPDA), Federal Rural University, Rio de Janeiro, where he lectures and researches on the global agri-food system and economic sociology. He has published widely in both areas. In 2013, he coordinated the report *Biofuels and Food Security*, HLPE/FCS/FAO, and in 2016, coedited *O Sabor da Origem* with Paulo Niederle and Gilberto Mascarenhas. In recent research, he has returned to work on radical innovations in agri-food (together with Ruth Rama), and with David Goodman, he has contributed to recent discussions on food regime approaches.

Introduction

Alessandro Bonanno, Kae Sekine and Hart N. Feuer

This volume is about the emancipatory power and democratizing role of geographical indication (GI). Specifically, and through the presentation of original international research, it probes whether the implementation of GI represents a progressive alternative to the socioeconomic trends and outcomes that characterize the contemporary global neoliberal agri-food system. Supported by a long-standing debate on its relevance and ability to affect agricultural markets, GI is defined as "a sign used on products that have a specific geographical origin and possess qualities or a reputation that are due to that origin" (WIPO 2015, 8). Moreover, it is often referred to as tool that attaches additional market value to a product by linking it to a specific place, its culture and traditional forms of production (Artini et al. 2016; Parasecoli 2017). The place-based value of GI products (in short, GIs) is further legitimized by the concept of terroir. Lacking adequate translation in many major languages, including English, terroir is used in its original French to indicate the combination of environmental, cultural and socioeconomic factors that shape the production of a specific agri-food product (Patterson and Buechsenstein 2018; Wilson and Johnson 1999). An agri-food product, this construct indicates, has a set of inalienable characteristics that distinguish it from, and make it more valuable than, similar products. The probing of the theme of the relevance GI as a tool for socioeconomic development and democratization of agri-food follows a sociological approach. This posture implies placing the analysis within the context of not only sociological debates – as opposed to legal and administrative discussions, as it is frequently the case – but also relevant social phenomena such the globalization of society and the neoliberalization of social relations. Additionally, it pays attention to the social implications of the evolution of market relations that characterize capitalism.

To be sure, the selection of the theme of the emancipatory role of GI should not be a surprise to those who follow the evolution of agriculture and food in the global neoliberal era. These students are aware of the many challenges that the corporatization of, and capital concentration in, agri-food engender for producers, consumers, agricultural communities and society as a whole (Bonanno and Wolf 2018; Howard 2016; Sekine and Bonanno 2016). In this context, relevant is the claim made by pertinent research about the desirable consequences engendered by the implementation of GI. Accompanying the assertion that GI adds value to local agri-food products, proposed arguments stress a number of additional beneficial consequences, including the augmentation of the economic well-being of producers and their communities, the availability of better quality food for consumers, the protection of heritage products and safeguarding the environment (Calboli and Wee Loom 2017; Paus 2010; Van Caenegem and Clearly 2017). However, GI is also a complex concept/process whose evolution contains conditions that are contradictory and, in some instances, inadequate to promote the well-being of stakeholders, their communities and, ultimately, society as a whole (Gangjee 2012; Guthman 2004; Hughes 2017; Parasecoli 2017; Zukin 2008). Given this background, the research question – whether the implementation of GI is a tool that can improve the conditions of all its stakeholders – earns further legitimacy.

The contradictory nature of GI finds its roots in the origin of the concept and its early implementations. As Blancaneaux (this volume) documents in the mature example of French wine, GI was implemented in the early 20th century to instill discipline among regional producers in order to uphold quality within a context of unstable agriculture (plant disease) and market shifts (increasing urban interest in certified wine), but the integrity of the system incrementally eroded as the number of certified regions reached saturation and as industrial-scale production became commonplace in GI regions. This points to one of the fundamental paradoxes of GI, namely that production regions are encouraged to maintain historical production systems and delimited borders but are also inexorably pushed by market forces to expand production and penetrate new regions. This integration of GIs into trade, particularly international trade, has thus broadly come to define the raison d'être of GI. The effort to reconcile this contradiction is particularly sharp in developing and agri-export-oriented countries such as Turkey (see Nizam, this volume) and Cambodia (Feuer, this volume). The origin of GI in Western Europe and its expansion abroad as a form of intellectual property protection for European products has thus been, on reflection of world systems theory, considered as tantamount to neocolonialism (Arewa 2006; Broude 2005; Pretorius 2002). In fact, even exporting the idea that Indigenous products and their associated intangible cultural capital should be commoditized and traded (even at higher value) is questioned (Frankel 2011; Posey 1990).

However, in the contemporary globalized neoliberal world, the rather frequent and heterogeneous use of the term *GI* perhaps best defines the complexity of its status. Neoliberal globalization has promoted the diffusion of local agri-food products worldwide (Bonanno and Constance 2008; Burch and Lawrence 2007; Howard 2016). This situation rests on the standardization of production and products and the opening of markets whereby the elimination of barriers to the circulation of commodities allows increasingly homogenous products to move with accelerated velocity across the socioeconomic space (Bonanno 2017). Supported by the neoliberal construct of unrestricted competition, the standardization of production and the increased global circulation of agri-food goods have fostered the emergence of *food from nowhere*. This phenomenon refers to the dominant status of food items whose *ubiquity*, *anonymity* and *uniformity* promote relatively low prices and high levels of availability and consumption. These are factors that, through the universalization of industrial food, augment the profits of large transnational corporations (TNCs) and their control of all facets of food production and distribution (Carolan 2013; Clapp and Fuchs 2009; Sekine and Bonanno 2016; Wolf and Bonanno 2014). Additionally, this situation fosters conditions that penalize family and Indigenous producers, labor, local farming communities, and the socioeconomic development of less-advanced regions, while undermining the safeguard of the environment and preservation of natural resources.

Opposition to the global neoliberal evolution of agri-food resulted in the emergence of not only incisive critiques of corporate agriculture but also practical initiatives that include civic agriculture, farmers markets, organic production, urban agriculture, slow food and more. Despite its pre-neoliberal origins, GI is often placed on this list of instruments designed to oppose corporate agri-food by privileging the stabilization of foods' origins, histories and traditions as prescribed by the concept of terroir (Calboli and Wee Loom 2017; Van Caenegem and Clearly 2017). Indeed, one of the main benefit of GIs is that they are impersonal and not connected to a specific rights holder. This means that first arrival or market position does not necessarily lead to the consolidation of power (Addor and Grazioli 2005, 870). However, because GI policies are inherently designed, and indeed promoted to third countries, as a malleable and culturally/institutionally adaptable framework (O'Connor and Company 2005), they are vulnerable to juridical and legislative loopholes that can be taken advantage of by politically or economically powerful agents. In practice, then, TNCs are in a position to co-opt various GIs if the product in question can be shaped to fit capitalist logic. Contributors to this book present cases across this spectrum: TNCs having low interest in a relevant GI (Sekine, this volume), TNCs deflected by public producers (Hegnes and Amilien, this volume), TNCs partially capturing a production region (Feuer, this volume) and TNCs aggressively appropriating a GI (Renard and Domínguez Arista, this volume). In great empirical detail, these cases demonstrate common tactics through which the "spirit" of GI can be undermined as well as some institutional mechanisms that are employed to preempt co-optation.

Due to the crisis of traditional forms of contestation such as strikes and demonstrations, almost all these other initiatives are based on individual and consumption-centered actions (Bonanno and Wolf 2018). These aspects played a relatively limited role in the historically relevant 20th-century process of producing cheap and abundant food for the expanding urban/industrial population (Clapp 2016; Le Blanc 1999). But as the social dimension of food evolved from its original role of a basic social need to the contemporary cultural experience, the objective of generating cheap and abundant food became not only less attractive but also a process that should be questioned and ultimately opposed (Carolan 2013; Parasecoli 2017).

Food as understood as a social experience entails a number of important social implications that remain relevant for the possible emancipatory dimension of GI. First, and as already mentioned, the issue of alternative agri-food is framed almost exclusively in individual terms. Accordingly, individuality and individual preferences take primacy over structural, collective factors – such as industrial relations, conditions of production and power concentration – in establishing desirable developments for agri-food (Warde 2014). The wishes of individual consumers, in other words, represent the ultimate factor of the way agri-food ought to be. Second and because individuals act as consumers, the evolution of agri-food remains exclusively contained within market relations. This implies that food can be understood only as a commodity as its raison d'être is centered on production for market exchange and the generation of profit. Alternative views of food – such as food as a human right or as a decommodified entity – cannot be contemplated. A particular contradiction of GI emerges in this, as specialty products are meant to be simultaneously exclusive, unique and rare, whereas specialty production (i.e., new GI production certification) is meant to be widely accessible to all producers as a form of rural development. Third, as market relations remain transcendental, the best possible option for the future organization of agri-food is that of a "fair" market. This outcome, however, remains based on a situation that offers inconsistent instruments to combat the tendency toward the concentration and centralization of resources typical of capitalism and the ability of large corporations to co-opt alternative forms of producing and distributing foods. Finally, individual food consumption understood as an experience is income-sensitive. This means that affluent segments of society are those that are more likely to be involved in the formal practices of this type of consumption. While this situation is not necessarily undesirable, the discrimination of less affluent social groups represents an indictment of this understanding of food.

In the specific case of GI, the alternative to the dominant form of corporate agri-food and the global neoliberal regime that supports it is expressed in at least three ways: (1) the defense of tradition and its cultural traits; (2) the requalification of the local with its uniqueness and history and (3) protection from the unwanted consequences of market competition. As far the issue of the *defense of tradition and its cultural traits* is concerned, particular importance is given to the disintegrating consequences of the domination of food from nowhere. As the production and consumption of industrial food damage small and medium producers and consumers alike while benefiting corporations, attempts to recover the cultural and traditional dimensions of food are seen as generalized antidotes against the standardization and mass production of food.

Largely dwelling on the concept of terroir and the importance of the local, the implementation of GI initiatives centers on efforts to differentiate traditional and culture-rich local foods from the poverty of food from nowhere. In this context, the homogenization and standardization promoted by neoliberal globalization are interpreted as factors that foster the disappearance of specific identities and ways of life as the local is colonized by the transcendental global.

Local food, therefore, is transformed into an instrument to oppose the otherwise so appealing predictability, affordability and convenience of industrial food and the loss of identity that goes with it. The familiarity of global food brands, fast food restaurants and the overall food from nowhere, in other words, is countered by the rediscovery of the originally local, its culture and history and the different identities that it entails.

This requalification of the local and its history parallels broader contemporary identity movements, whereby difference is elevated as a source of strength and a proven form of emancipation. Paradoxically, however, this requalification of local food and its history stands not only on the phenomenon of globalization but also on the existence of global markets capable of absorbing the unique niche goods that the many "local" producers generate. At the cultural level, localism remains anchored in cosmopolitanism, revealing the dialectical relation between the local and the global and its social construction. As the global is understood as the outcome of transnational economic forces, the local with its tradition and history is also reified or invented to promote its historical agri-social dimensions that can be consumed in the market. The frame for the commodification of these "narrative" features of agri-food products (as opposed to only flavor, quality and aesthetics) is enabled by GI legislation. Inspired by a "modernist" search for an absolute "authenticity," the requalification of the local and its tradition actually manifests itself through highly negotiated and contested definitions that are "validated" by the varying opinions of experts and magistrates and deny the dynamic and contingent histories that led to the contemporary "authentic."

Emphasis on the local further translates into providing importance and support to established local ways of life, ways of production and their products. Because of their uniqueness, these established practices are viewed as threatened by not only modernization but also any form of progress. In essence, this posture offers a rejection of the proposition that progress could add to the quality of local life, production and products, assuming, by definition, that the past is more desirable than the future. As this might be the case in numerous situations in which modernization led to the colonization and/or destruction of the local, this view further proposes that a conservation of the past is a desirable response to the contractions of the present and the uncertainties of the future. Glossing over the many contradictions associated with the unconditional support for the "desirability" of the past, this posture oddly equates conservativism with progressiveness.

The social construction of the local that is typical of GI unconditionally presumes the natural superiority of the local over the global. Local food is always assumed – rather than proved – as better than other types of food by valorizing dimensions, such as tradition and small scale, which are socially constructed (tradition and locality make it better, by definition). Additionally, it is automatically endowed with desirable characteristics, such as natural taste, and often unjustifiably associated with positive outcomes, such as the safeguarding of biodiversity and the environment. Although it is undeniable that some of these aspects would survive the rigorous scrutiny of critics for many products, it is also the case that these contentions distract from the more fundamental limits of fetishizing the local and the fallacy of automatically equating the local with better. As documented by research presented in this volume and elsewhere, local traditions and history are not immune from undesirable social phenomena such as racism, discrimination, labor exploitation and inequality.

The unintended or unexpected outcomes of GI policies, what we refer to as policy contradictions, arise due to the disparity between the wide range of positive expectations of GI and the limited number of characteristics that are commonly, or can legally be, protected. This reflects the conflation of GI as an ideal (a historical regional product) and as a legal tool (a certified product), a reality which is coming into starker relief as more countries in the world adopt discrete GI policies. In many of these new countries, the technical processes of setting up a mechanism for certification takes a back seat to the popular imagination of high-value exports and aspirations for domestically famous products to join the rarefied space occupied by Champagne, Parmigiano Reggiano and Jamón serrano. But the realities of competing within the high-value agri-food market place dominated by Europe may finally represent only an illusory opportunity to rectify chauvinistic cultural hierarchies of food culture established through imperialism (Arewa 2006). Other common misconceptions that are not exclusive to new-entry GI countries include unjustified assumptions about the prevalence of secondary impacts, such as safeguarding the environment and protecting small-scale farmhouse production. This confusion surrounding the GI ideal and its technical implementation arises in the debate about the protection of Indigenous knowledge, which is often highlighted as one of the important benefits of GI without critically reflecting on whether, or to what extent, this encodes repressive or patriarchal practices (Coombe and Malik 2017).

Despite the documentation of such contradictions by legal scholars and rural sociologists, the notion of embeddedness remains a fundamental component of assumed opposition to the global neoliberal food system. It is the unique way in which food is locally produced and prepared that allows local communities to claim ownership over this component of their ways of life and to oppose banal consumerism and homogenization. Accordingly, processes that oppose the global neoliberal colonization of the sphere of food production and consumption cannot fail to consider the importance of embeddedness. It is the relevance of scrutinizing the importance of the local that further legitimizes the objective of this book.

The potential use of GI as a *protection against unwanted consequences of market competition* places it among the tools designed to steer the economy away not only from crises but also from undesirable patterns of development. Additionally, it is seen as an instrument that opposes colonizing corporate forces and their ability to co-opt local tradition, culture and history. The global-local dialectical relationship previously mentioned is relevant here as well. It allows GI foods to access expanding global markets. Yet simultaneously, it provides them the necessary protection to mitigate the consequences of open competition. To achieve this situation, culture, history and tradition are transformed into commercial items that add value to these food items. Following a classical sociological analysis, there is a requalification of culture, history and tradition from entities based on substantive values to entities that find their power through their commercialization. Accordingly – and as these are socially constructed items – their construction takes an *instrumentally rational* turn, as it needs to be commensurate to the requirements of the pursuit of profit. Alternative forms of rationality, such as self-sufficiency, barter networks, informal quality approval and other non-commoditized forms of agri-food circulation often referred to as the "culture economy," are therefore automatically excluded (Ray 1998). Obviously, this is not uncommon; it is a dominant occurrence under neoliberalism. However, it signals the fact that alternatives to the functioning of the free market are constructed in market terms.

Unrestricted competition and market fluctuations remain among the most significant dangers that GI proponents wish to oppose. In this context, the protection of local products that the GI label offers is accompanied and, in some instances, substituted by cases in which the value of local products, productions and ways of life is not simply exported but employed to requalify the local and attract the business of non-local actors. Specifically, this situation refers to initiatives, such agritourism, local culinary festivals and celebrations of the local food and landscape, that transform the local not only into a tool for economic development but also into an instrument that empowers Indigenous communities and promotes the safeguarding of biodiversity and the environment.

This reality suggests why there are no systematic and structural safeguards to ensure the emancipatory power of GI, even as there are opportunities to mitigate or defang some of the undesirable consequences of capitalism in some cases. In this volume, we explore the future viability and usability of GI as it spreads to new territories and matures in its core regions. In each chapter, authors engage with the expectations, assumptions and contradictions of GI while contributing to a broader dilemma: Can the emancipatory elements of GI be obtained by tinkering and optimizing the legislative apparatus, or will GI wage a losing campaign against the inexorable forces of neoliberalism? To this end, the case studies reflect on the constellation of institutional and juridical factors that impact the capacity for each region or country to meaningfully challenge the neoliberal co-optation of some historically important or symbolic products.

The evolution of GI

The roots of the GI system date back to the BCE era (Takahashi 2015). In ancient Greece, for instance, the terms *Corinthian wine* and *almond of Naxos* associated these products' qualities with their places of origin. In the Middle Ages, the seals of guilds indicated the quality of locally produced food items (Thévenod-Mottet and Marie-Vivien 2011). As markets expanded and the distance between producers and consumers increased, geographical identification – often in the form of recorded names and/or logos – became increasingly sought after to communicate agri-food products' quality, features and reputations to consumers. With expanding use, pertinent legal measures were introduced. In the early modern era, legislations concerning GIs appeared in France (e.g., Roquefort in 1411 and Bresse Chicken in 1591), Switzerland (e.g., Gruyere in the 17th century), Hungary (e.g., Tokaji-Hegalja wines) and Italy (e.g., Chianti) (Gatti, Giraud-Héraud, and Mili 2003; Takahashi 2015).

In the 18th century, the further expansion of agri-food trade prompted the signing of international agreements to protect well-known agri-food items, as in the case of the 1712 trade agreement that granted protection to France's Champagne, Bordeaux wines and Provence olive oil (Takahashi 2015). In 1883, the signing of the *Paris Convention for the Protection of Industrial Property* marked the establishment of the first GI international accord (Calboli and Wee Loon 2017; Echols 2008; Parasecoli 2017; Takahashi 2015; Thévenod-Mottet and Marie-Vivien 2011). The primary purpose of this document was to prevent fraud, consumer deception and unfair competition. Still in effect, the Paris Convention, currently counts on 177 signatory countries (World Intellectual Property Organization 2018a). The following 1891 Madrid Agreement provided specific rules for indicating a food item's source. In 2018, it counts the participation of 101 countries (World Intellectual Property Organization 2018b).

In the early 20th century, agri-food product quality deterioration and fraud prompted the introduction of legislation in a number of European countries. In France, measures to prevent consumer misinformation about wines' quality and origin were introduced in 1905. Similar legal procedures to protect additional agri-food items such as cheese, ham, olive, fruits and vegetables were instituted in 1919 (Calboli and Wee Loon 2017; Echols 2008; Parasecoli 2017; Takahashi 2015; Thévenod-Mottet and Marie-Vivien 2011). In Spain, legislation for the protection of local wines dates back to 1925, whereas in Italy – the country with the second greatest number of GI agri-food products – similar rules were introduced in the post–World War II years. Simultaneously, the rapid expansion of agri-food commerce led to new international agreements such as the 1951 *Stresa Convention* on cheese that was signed by eight countries and the 1958 *Lisbon Agreement* that, signed by twenty-seven countries, offered international protection to agri-food products and the registration of appellations of origins. These agreements sanctioned GI's global role as a tool that identifies agri-food products and their quality.

When the GATT Uruguay Round (1986–1994) and the subsequent World Trade Organization (WTO) regime (1995–present) radically reduced tariff barriers and price support programs for agri-food products, the European Union and its allies engineered a new role for GI. These countries understood GI as a tool to support not only traditional and disadvantaged forms of agriculture – including family farming and family farms located in mountain regions and islands – but also rural communities and traditional cultures (Calboli and Wee Loon 2017; Sekine 2015). Globalization, urbanization and the industrialization of agri-food and their negative consequences on rural communities also provided shared support for this position (Parasecoli 2017).

At the international level and within the framework established by the WTO system, the application of GI legislation turned into a controversy between the European Union (EU) and the United States. The European Union and countries that adopt its model – also known as "Old World Countries" or "GI-friends" – share the tenet that the quality, characteristics and reputation of GI products are strictly confined to their places of origin and therefore cannot be reproduced in other locations. Subscribing to the concept of terroir, these countries favor a GI system that grants a high level of protection to agri-food items with GI status. Known as sui generis, this system consists of three categories of protection. The first, protected designation of origin (PDO), is granted to products that are uniquely and exclusively associated to a geographical area. In this case, production, ingredients, labor and know-how should be generated in the designated area. The second category, protected geographical indication (PGI), is less stringent in that it allows for ingredients to originate outside the designated area. The third and even less stringent category, traditional specialties guaranteed (TSG), refers to products that are not associated with a specific area but instead require the use of traditional ingredients and production know-how (Calboli and Wee Loon 2017; Echols 2008; Ilbert 2012, Parasecoli 2017; Takahashi 2015; Thévenod-Mottet and Marie-Vivien 2011).

The United States and a host of additional countries – including Australia, Argentina, Canada and Chile, also known as New World countries or as the United States and the Cairns Group – employ the mark system, wherein GI status can be granted to protect the commercial value and uniqueness of products. In this case, GI items can be freely produced by those who own the appropriate rights. Accordingly, rather than the association with a place, its culture and history (i.e., terroir), this system emphasizes brand recognition since GI agri-food products are distinguished by the quality and reputation of their brand names. The mark system consists of two categories: *trademarks* and *collective marks*. The trademark category refers to granting to an individual or a company the exclusive rights to produce, use and exploit a specific product. This product is considered private property, and its production is at the exclusive discretion of its owner. Collective marks indicate goods whose production is associated with the activities of a group that owns the mark. The group determines the standards and quality control of the mark. The coexistence of, and competition between, these two GI systems continues as the European Union and the United States ask uncommitted countries to support their own system, switch or apply a dual system (Parasecoli 2017; Sekine 2015).

This international dispute has unfolded despite the implementation of the 1995 TRIPS Agreement (Trade-Related Aspects of Intellectual Property Rights Agreement). This is an international treaty that was designed to protect GIs and includes among its signatory countries not only the European Union and the United States but also almost all the other countries in the world. In 2003, the United States and its allies challenged the EU GI legislation in the

WTO, citing that the European Union was in violation of TRIPS (Echols 2008; Parasecoli 2017; Sekine and Bonanno 2018; Takahashi 2015). In 2005, the WTO rejected the US claim, concluding that the European Union's GI legislation does not violate the terms imposed by TRIPS. However, because of the long-standing negotiations and concessions between the two sides, the protection of GIs within TRIPS remains open to both systems. First and as claimed by the European Union, TRIPS grants the highest level of protection only to wine and spirits. Simultaneously and as claimed by the United States, it accords a lower level of protection to other agri-food products (Thévenod-Mottet and Marie-Vivien 2011). Second, TRIPS allows signatory countries to independently establish GI regulation and, in effect, de facto permitting the coexistence of the sui generis and mark systems. Additionally, it allows host country courts to adjudicate disputes over GIs. Finally, the TRIPS Agreement's definition of the interconnection among the quality, characteristics and terroir of GI products is less stringent than that contemplated by the Lisbon Agreement (Calboli and Wee Loon 2017; Thévenod-Mottet and Marie-Vivien 2011). The Lisbon Agreement indicates that "The country of origin is the country whose name, or the country in which is situated the region or locality whose name, constitutes the appellation of origin which has given the product its reputation" (Article 2[2]). It also stipulates that "'appellation of origin' means the geographical denomination of a country, region, or locality, which serves to designate a product originating therein, the quality or characteristics of which are due *exclusively or essentially* to the geographical environment, including natural and human factors" (Article 2[1]). Arising from this strict view, the Lisbon Agreement has been contracted after 60 years by only 29 countries. Conversely, the TRIPS Agreement employs a more permissive understanding of GI, whereby "a given quality, reputation or other characteristics of the good are *essentially* attributable to its geographical origin" (Article 22[1]). In essence, it endorses a notion of GI that combines two of the aforementioned categories of GI contemplated by the EU system: PDO and PGI. Following the notion adopted by the TRIPS Agreement and the European Union's PGI certification, GI status can be awarded to products whose reputation is simply linked to the pertinent geographical area but whose ingredients are not local.

In the new century, the consequences of the 2005 WTO ruling and the lenient nature of the TRIPS definition of GI translated into the worldwide proliferation of lessened forms of sui generis GI legislations. This notably includes a number of newcomer countries in East Asia, such as Malaysia, Indonesia, Cambodia, Thailand, Vietnam, China, Japan, South Korea and even North Korea. In 2015 and following the TRIPS Agreement and the European Union's GI legislation, the Genova Act amended Article 2(1) of the Lisbon Agreement (Calboli and Wee Loon 2017; Thévenod-Mottet and Marie-Vivien 2011). This revision identifies two categories of GI: geographical denomination – that is, equivalent to the European Union's PDO – and geographical indications – that is, equivalent to the European Union's PGI. It also contemplates the existence of GI products less connected with their origins, such as TSG. In its effort to promote the adoption of the sui generis approach by other countries, the European Union supported the Genova Act. However, the support of reduced forms of sui generis GIs can be interpreted as a contradictory step that limits the very use of the concept of terroir in the granting of GI status in the attempt to expand the use of the sui generis approach. Additionally, the less stringent approach to the granting of GIs is generally considered more favorable to large agri-food corporations than to small and medium producers, further contradicting the claim that GI is a tool that promotes local, artisanal and small-scale production.

The organization of the book

The book's research question, whether GI offers a progressive alternative to the socioeconomic trends and outcomes that characterize the contemporary global neoliberal agri-food system, is explored through twelve original chapters and a conclusion, which are organized into four parts. Chapter 1 – the only chapter constituting the first part of the book – "Theoretical assumptions" – is authored by Alessandro Bonanno: "Geographical Indication in agri-food and its role in the global neoliberal era: a theoretical analysis" discusses the role of GI under the global neoliberal agri-food regime. Bonanno's argument rests on the analysis of two opposite theories that define the debate on GI. The first is the dominant neoliberal theory of the free market, whereas the second refers to the broad group of theories that contemplate the political and state-based regulation of the market. Under each of these two theories, the foreseen role of GI differs in order to correspond with the various formulae of socioeconomic development. Stressing the characteristics and contradictions of these two theories, Bonanno contends that in the neoliberal camp, the support of intellectual property rights contradicts the fundamental neoliberal tenets of the open access to markets and unrestricted competition. In the state intervention camp, the state regulation of markets promotes conditions that contradict the requirements for market expansion. These two sets of contradictions, Bonanno concludes, problematize the use of GI as a tool for the emancipation of subordinate groups.

The second part of the book is entitled "The Asian context" and includes chapters that directly discuss instances of the implementation of GI policies in Asia. The first of these chapters, by Hart N. Feuer, is "Geographical Indications out of context and in vogue: the awkward embrace of European heritage agricultural protections in Asia." Feuer sheds light on the rollout of European-style GI laws in East Asia by comparing Japan and Cambodia. He documents how Asian countries share a perception of terroir related to agri-food products which has, in addition to pressure from the WTO, animated almost all countries to adopt sui generis GI policies. In both of the evaluated countries, Feuer determines that the institutional models of GI ultimately adopted are broadly in line with European norms, but they more quietly emphasize the domestically important secondary benefits of GI, such as public quality assurance, rural development and marketing. This strategic, utilitarian deployment of GI policies may, however, hasten the decline in consumer's independent capacity for evaluating food quality and threatens the undocumented diversity associated with small-scale decentralized production and informal quality evaluation.

In the following chapter, "The Impact of geographical indications on the power relations between producers and agri-food corporations: a case of powdered green tea matcha in Japan," Kae Sekine discusses the extent to which GI contributes to the democratization of existing power relations in the food system and promotes sustainable territorial development. Her analysis is centered on the case of the Nishio matcha, which is a GI certified under the collective trademark and sui generis systems. Both these systems are currently employed in Japan. Sekine offers a brief history of the matcha (powdered green tea) food system, stressing its structure and changes. She also underscores the conflicting interests of the two primary socioeconomic groups characterizing this production. Small family producers of green tea leaves operate in a market controlled by processing corporations who set prices, grade quality and shape production practices. This situation leaves family producers in a subordinate position to corporate processors, who then pass on this economic vulnerability to tea leaf pickers as poorly paid manual labor. Given these conditions, Sekine concludes, this case study shows that GI does not systematically guarantee the democratization of the agri-food system and ensure sustainable territorial development. Accordingly, corrective public policies for the implementation of GI should be implemented.

This book's second part concludes with the chapter "Provenance for whom? A comparative analysis of geographical indications in the European Union and Indonesia," by Cinzia Piatti and Angga Dwiartama. Comparing the significantly different GI systems of Indonesia and the European Union, the authors propose a three-dimensional analysis – based on the scrutiny of sociocultural factors, the ecological context and power relations – that allows them to illustrate the three moments characterizing the creation of GIs. These moments consist of market integration, institutionalization and the acquisition of provenance labels. By analyzing these moments, Piatti and Dwiartama stress how power differentials are created and eventually manifest themselves and the implications that this process has for the formation of alternatives to the dominant neoliberal agri-food regime. Their conclusions underscore the problematic nature of GI as an emancipatory tool. They stress that while dominant social and power relations are reproduced despite GI processes, the possibility – albeit more theoretical than actual – for different patterns of socioeconomic development continues to exist.

The third part of the book – "Cases from Europe" – begins with a chapter by Derya Nizam: "How to use geographical indication for the democratization of agricultural production: a comparative analysis of geographical indication rent-seeking strategies in Turkey." In line with the volume's research question, Nizam's objective is to analyze the potential of GI to generate local institutional resources that can empower agri-food producers and rural communities in a way that helps them to combat their dependence on the dominant agri-food regime. Analyzing the cases of three olive oil GIs from Turkey's Aegean region by using the commodity chain analysis method, Nizam illustrates the characteristics of these GI products and tracks their respective processes of formation and governance. In her analysis of the diverging outcomes among the cases, she concludes that participatory forms of governance (reflexive localism) offer better opportunities for more equitable and independent forms of socioeconomic development. These forms of governance do not depend on GI but may be able to use the process of GI formation as a supportive context.

The following chapter, "Geographical indications – a double-edged tool for food democracy: the cases of the Norwegian geographical indication evolution and the protection of stockfish from Lofoten as cultural adaptation work," by Atle Wehn Hegnes and Virginie Amilien, discusses the importance of the process of adaptation in the implementation of GIs. Probing the overall question whether GI is a universal tool for local socioeconomic development, the authors analyze the Norwegian case of stockfish from Lofoten. They argue that the abstract construct of GI needs to be translated into practical forms of adaptation. These forms of adaptation are essential components of the implementation process. Additionally, they continue, adaptation work is fundamental because the results of GI implementation heavily depend on how social, cultural, political and economic interactions take place, including the interface between the complex local sphere and the global ordering established by WTO regulation. Concluding that GI is not a universal tool for socioeconomic development, the authors stress not only the complexity of the consequences that adaptation entails but also, and more importantly, the emergence of contradictory conditions that cause adaptation and simultaneously promote processes of aggregation, emancipation as well as alienation and exclusion.

Adding to the discussion on instances from Europe, the chapter by Romain Blancaneaux, "The decline of the French label of origin wine," probes the consequences of the proliferation of GIs in the wine sector in France. Blancaneaux illustrates that the original emergence of the denomination of origin (DO) labels provided an effective form of protection for producers and consumers and a way to safeguard the quality of wines and stabilize markets. Over time, however, the market for DO wines swung between two extremes: exclusivity and profusion. In the early decades, production restrictions and quality standards increased prices, making protected wines accessible only to affluent consumers. As the number and production of DO wines increased, the exceptionality of protected wines was gradually transformed into the norm, de facto eliminating many of the advantages that the early implementation of this process had generated. As a result, the assumption that GI is a tool that can be increasingly refined over time to better realize socioeconomic development has been strongly discredited. The chapter concludes by cautioning against optimistic assumptions about the increasing abundance and long-term use of GIs.

The chapter by Anastasiya Shtaltovna and Hart N. Feuer "Modern resilience of Georgian wine: geographical indications and international exposure" concludes this section. Using wines from the Republic of Georgia as an example,

Shtaltovna and Feuer describe the resilience that winemaking has achieved by using idiosyncratic social and political processes. Considered the oldest winemaking region of the world and having survived a great number of political and economic changes over thousands of years, Georgia and its winemakers have been able to defend and maintain unique varieties and production specifications that are embedded in local culture and history. This resilience, the authors continue, is associated with broad support from elite and peasant groups alike. Pushed by neoliberalism and enhanced exposure to international wine conventions, the government is hoping to employ GI to perpetuate this resilience of their wine culture. In this effort, GI is understood as a potential tool to valorize and protect tradition, but given their past successes in using domestic resources, local producers are wary of external mechanisms. Shtaltovna and Feuer observe that there are important differences between the European-created concept of GI and the Georgian wine tradition, in that the Georgian understanding of quality in wine typically privileges varietal diversity over territory and is co-created through consumption patterns that are embedded in historical social networks. The potential discrepancy with the European conceptualization of quality wines leads Shtaltovna and Feuer to question whether GI can realistically form a parallel track of valuation for export that does not disrupt the Georgian tradition.

The fourth and final part of the book is devoted to cases from the Americas. It opens with two chapters, each of which discusses cases from Latin America. The first of them is "The multilevel, multi-actor and multifunctional system of geographical indications in Brazil," by Paulo Niederle, John Wilkinson and Gilberto Mascarenhas. Analyzing the case of Brazil, the authors contend that the understanding of GI as an emancipatory tool is shared in Latin America. GI is seen as an initiative that fosters the connection between producers and consumers by promoting short value chains and valorizing food culture and local know-how. Following this view, GI is considered part of the broader process of relocalization and patrimonialization of the food system. Simultaneously, they contend, GI products are open to co-optation by mainstream producers that use the institutional unevenness of GI to secure competitive advantages. Looking at three sectors – wine, coffee and cheese – Niederle, Wilkinson and Mascarenhas illustrate the advantages and disadvantages of the institutional flexibility of the Brazilian GI system. This flexibility is designed to respond to the different realities of the Brazilian productive and cultural landscape. However, it is also the result of the low levels of normative enforcement and the deliberate effort to transfer a significant part of decision-making to sectorial and territorial actors. In this context, institutional flexibility managed to adapt GIs to different realities. Simultaneously, however, it also creates openings for strategic behavior that can foster institutional arrangements to regulate the system. The net result, they contend, has been the creation of GIs that are based not on common principles but rather on the relative negotiation power of different stakeholders. They conclude that this complex and multifaceted GI system has engendered an institutional instability that diminishes its effectiveness as an emancipatory tool.

This part continues with the chapter "The geographical indication of mezcal in Mexico: a tool of exclusion for small producers," by Marie-Christine Renard and David Rodolfo Domínguez Arista. Employing the case of the GI status granted to mezcal in Mexico, Renard and Domínguez Arista illustrate the discriminatory consequences that this process generated. Mezcal, the authors document, is a traditional alcoholic beverage often distilled by small producers in various Mexican states. To promote its production and marketability and safeguard its tradition, the government of Mexico granted mezcal a denomination of origin label in the mid 1990s. Led by the industry elite of large producers, this process engendered the establishment of GI areas that followed established political-administrative borders rather than agroecological and historical boundaries. The result was the exclusion of a great number of small producers from protected zones. Attempts were made to force these producers to abandon the use of the word *mezcal* and instead to employ obscure labels, such as *komil* or *agave aguardiente*. In their conclusions, Renard and Domínguez Arista stress that this case is an instance in which GI is transformed from an emancipatory tool into an instrument of exclusion.

The next two chapters analyze instances from North America. The first of them is Anelyse M. Weiler's contribution: "Whose labor counts as craft? Terroir and farm workers in North American craft cider." Weiler illustrates the development of the niche market of craft cider in Canada and the United States. This market's projected image, Weiler contends, is constructed on the idea of small- to medium-scale artisans producing in a unique geographical region and following traditional practices and values. Some members of this network of producers, she adds, wish to enhance their commercial viability by establishing a GI. While still on the early stages of development, Weiler documents how the establishment of GIs raises a contradictory situation in which artisanal production rests on the widespread employment of racialized immigrant labor even though this phenomenon clashes with the ideals of many producers and consumers. In effect, Weiler continues, the success of artisanal cider production is based on the exploitation of immigrant farm workers that are deprived of the material and symbolic rewards associated with this production. This is a situation, she concludes, that is difficult to rectify, because it requires the elimination of significant logistical and cultural barriers through the introduction of extra-sectorial policy changes.

The second and last chapter of this final part is "The potential role of geographical indication in supporting Indigenous communities in Canada," by Donna Appavoo and Monika Korzun. It probes the issue of whether GI can be a beneficial tool for Indigenous communities and a source of protection for Indigenous food in Canada. Claiming that this is an important and yet understudied topic, Appavoo and Korzun illustrate the similarities between the classic objectives of GI and the challenges that Indigenous people face. These challenges include the protection of Indigenous knowledge; the safeguarding of history; the defense of traditional ways of life, including food and food sources; the expansion of rural employment; and the fortification of rural alliances. They also discuss the shortcomings of GI, stressing that it can marginalize or even penalize family producers and communities, reproduce power unbalances and allow corporations to use it to their advantage. Analyzing a number of key aspects relevant to GI schemes and Indigenous communities – such as local knowledge, land access, market access and legal protection – Appavoo and Korzun conclude that despite the possible limits of GI, it has a potentially important role to play in the emancipation of Indigenous communities in Canada.

The volume ends with "Conclusions: comprehensive change and the limits and power of sectorial measures," by the book editors. Based on, but also adding to, the arguments presented in the volume, it provides an answer to the book's research question about the emancipatory power of GI under globalization and neoliberalism. It opens with a brief summary of the overall conclusions derived from the book, which critically reflect on the potential for GI to transform and re-organize the agri-food sector in spite of broader constraints at the societal level. Overcoming thorny agri-food constraints requires comprehensive approaches that involve the restructuring of current social arrangements. Following this analysis, the overarching conclusion presented is that GI does not offer systematic solutions to structural problems highlighted by some classic and contemporary literature. However, while GI cannot structurally change the agri-food sector, it generates a number of consequences that can be relevant for emancipatory efforts. In this regard, the chapter indicates six additional conclusions. First, it concludes that the emancipatory role of GI depends on local social relations and socioeconomic conditions, which are in turn contingent on existing structural factors and the exercise of human agency. The second conclusion states that the success of GI depends on the way it is implemented. Implementation involves contradictory processes that may include the loss of heterogeneity, local identity and vernacular institutions. The third conclusion states that GI does not necessarily represent a safeguard against the functioning of the free market, because emancipatory results are often endogenously countered by opposing consequences. The fourth conclusion underscores that GI represents a socially and politically aggregating force that is both necessary and beneficial but is often accompanied by episodes of distrust and conflict. The fifth conclusion refers to the claim that GI protects the environment and supports the sustainable use of natural resources. This claim can be rejected because historical food systems upon which GIs are based do not consistently inhere ecological values. The sixth conclusion indicates that GI supports the desire of producers to remain in farming and food production despite adverse conditions. The chapter ends by contending that the emancipatory power of GI in all its implementations is contradictory in that it contains elements that promote but also hamper its ability to improve the socioeconomic conditions of stakeholders and democratize the agri-food sector.

Bibliography

Addor, Felix, and Alexandra Grazioli. 2005. "Geographical Indications Beyond Wines and Spirits: A Roadmap for a Better Protection for Geographical Indications in the WTO/TRIPS Agreement." *The Journal of World Intellectual Property* 5 (6): 865–97.

Alkon, Alison Hope, and Julian Agyeman. 2011. *Cultivating Food Justice: Race, Class, and Sustainability*. Cambridge, MA: MIT Press.

Arewa, Olufunmilayo B. 2006. "TRIPs and Traditional Knowledge: Local Communities, Local Knowledge, and Global Intellectual Property Frameworks." *Marquette Intellectual Property Law Review*10 (2): 156–82.

Artini, Filippo, Maria Cecilia Mancini, Mario Veneziani, and Michele Donati, eds. 2016. *Intellectual Property Rights of Geographical Indications*. Newcastle upon Tyne: Cambridge Scholars Publishing.

Bonanno, Alessandro. 2017. *The Legitimation Crisis of Neoliberalism*. New York: Palgrave Macmillan.

Bonanno, Alessandro, and Douglas H. Constance. 2008. *Stories of Globalization*. University Park, PA: Pennsylvania State University Press.

Bonanno, Alessandro, and Steven A. Wolf, eds. 2018. *Resistance to the Neoliberal Agri-Food Regime: A Critical Analysis*. New York: Routledge.

Broude, Tomer. 2005. "Taking 'Trade and Culture' Seriously: Geographical Indications and Cultural Protection in WTO Law." *University of Pennsylvania Journal of International Economic Law* 26 (4): 623–92.

Burch, David, and Geoffrey Lawrence, eds. 2007. *Supermarkets and Agri-Food Supply Chains: Transformations in the Production and Consumption of Food*. Cheltenham, UK: Edward Elgar Publishing.

Calboli, Irene, and Ng-Loy Wee Loon, eds. 2017. *Geographical Indications at the Crossroads of Trade, Development, and Culture*. Cambridge: Cambridge University Press.

Carolan, Michael. 2013. *Reclaiming Food Security*. New York: Routledge.

Clapp, Jennifer. 2016. *Food*. Malden, MA: Polity Press.

Clapp, Jennifer, and Doris Fuchs. 2009. *Corporate Power in Global Agri-Food Governance*. Cambridge, MA: MIT Press.

Coombe, Rosemary J., and Ali Malik. 2017. "Rethinking the Work of Geographical Indications in Asia: Addressing Hidden Geographies of Gendered Labor." In *Geographical Indications at the Crossroads of Trade, Development, and Culture: Focus on Asia-Pacific*, 87–121. Cambridge: Cambridge University Press.

Echols, Marsha A. 2008. *Geographical Indications for Food Products: International Legal and Regulatory Perspectives*. Alphen on the Rhine, The Netherlands: Kluwer Law International.

Frankel, Susy. 2011. "Mismatch of Geographical Indications and Innovative Traditional Knowledge." *Prometheus* 29 (3): 253–67.

Gangjee, Dev. 2012. *Relocating the Law of Geographical Indications*. Cambridge: Cambridge University Press.

Gatti, Silvia, Eric Giraud-Héraud, and Samir Mili. 2003. *Wine in the Old World: New Risks and Opportunities*. Milan: Franco Angeli.

Guthman, Julie. 2004. *Agrarian Dreams: The Paradox of Organic Farming in California*. Berkeley: University of California Press.

Howard, Philip. 2016. *Concentration and Power in the Food System*. New York: Bloomsbury.

Hughes, Justin. 2017. "The Limited Promise of Geographical Indications for Farmers in Developing Countries." In *Geographical Indications at the Crossroads of Trade, Development, and Culture*, edited by Irene Calboli and Ng-Loy Ween Loom, 61–86. Cambridge: Cambridge University Press.

Ilbert, Hélène. 2012. "Products with Denominations of Origin and Intellectual Property Rights: The International Bargaining Process." In *Geographical Indications and International Agricultural Trade: The Challenges for Asia*, edited by Augustin-Jean Louis, Hélène Ilbert and Neantro Saavedra-Rivano, 91–116. New York: Palgrave Macmillan.

Le Blanc, Paul. 1999. *A Short History of the U.S. Working Class*. New York: Humanities Books.

O'Connor and Company. 2005. *Geographical Indications and the Challenges for ACP Countries*. Wageningen, The Netherlands: Agritrade – Technical Centre for Agricultural and Rural Cooperation (ACP-EU).

Parasecoli, Fabio. 2017. *Knowing Where it Comes from: Labeling Traditional Food to Compete in a Global Market*. Ames, AI: Iowa State University Press.

Patterson, Tim, and John Buechsenstein, eds. 2018. *Wine and Place: A Terroir Reader*. Berkeley. University of California Press.

Paus, Marguerite. 2010. *Geographical Indications and Rural Development*. Saarbrücken: Verlag.

Posey, Darrell. 1990. "Intellectual Property Rights: And Just Compensation for Indigenous Knowledge." *Anthropology Today* 6 (4): 13.

Pretorius, Willem. 2002. "TRIPS and Developing Countries: How Level Is the Playing Field?" In *Global Intellectual Property Rights: Knowledge, Access and Development*, edited by Peter Drahos and Ruth Mayne, 183–97. London: Palgrave Macmillan.

Ray, Christopher. 1998. "Culture, Intellectual Property and Territorial Rural Development." *Sociologia Ruralis* 38 (1): 3–20.

Sekine, Kae. 2015. "What Are the Roles of Geographical Indication System in Japan?" *Agriculture and Economy* 81 (12): 62–70 (in Japanese).

Sekine, Kae, and Alessandro Bonanno. 2018. "Geographical Indication and Resistance in Global Agri-food." In *Resistance to the Neoliberal Agri-Food Regime: A Critical Analysis*, edited by Alessandro Bonanno and Steven A. Wolf, 106–19. London: Routledge.

———. 2016. *The Contradictions of Neoliberal Agri-Food*. Morgantown: West Virginia University Press.

Takahashi, Teiji. 2015. *Geographical Indications for Agri-food Products: A Guide on the GI System to Increase the Value of Regional Products*. Tokyo: Nobunkyo (in Japanese).

Thévenod-Mottet, Erik, and Delphine Marie-Vivien. 2011. "Legal Debates Surrounding Geographical Indications." In *Labels of Origin for Food: Local Development, Global Recognition*, edited by Elizabeth Barham and Bertil Sylvander, 13–28. Oxford: CABI International.

Van Caenegem, William, and Jen Clearly, eds. 2017. *The Importance of Place: Geographical Indications as a Tool for Local and Regional Development*. Cham, Switzerland: Springer.

Warde, Alan. 2014. "After Taste: Culture, Consumption and Theories of Practice." *Journal of Consumer Culture* 14 (3): 279–303.

Wilson, James E., and Hugh Johnson. 1999. *Terroir: The Role of Geology, Climate, and Culture in the Making of French Wines*. Berkeley. University of California Press.

Winders, Bill. 2017. *Grains*. Malden, MA: Polity Press.

WIPO (World Intellectual Property Organization). 2018a. *WIPO Administrated Treaties on the Paris Convention*. Geneva: WIPO. Online Document. Accessed September 11, 2018. www.wipo.int/treaties/en/ShowResults.jsp?treaty_id=2.

———. 2018b. *WIPO Administrated Treaties on the Madrid Protocol*. Geneva: WIPO. Online Document. Accessed September 11, 2018. www.wipo.int/treaties/en/ShowResults.jsp?treaty_id=8.

———. 2015. *Geographical Indications: An Introduction*. Geneva: WIPO. Online Document. Accessed June 6, 2016. www.wipo.int/edocs/pubdocs/en/geographical/952/wipo_pub_952.pdf.

Wolf Steven A., and Alessandro Bonanno. 2014. *The Neoliberal Regime in the Agri-Food Sector. Crisis, Resilience and Restructuring*. New York: Routledge.

Zukin, Sharon. 2008. "Consuming Authenticity: From Outpost of Difference to Means of Exclusion." *Cultural Studies* 22 (5): 724–48.

Part I

Theoretical assumptions

1 Geographical indication in agri-food and its role in the global neoliberal era

A theoretical analysis

Alessandro Bonanno

Introduction

This book on geographical indication (GI) probes the relevance of GI as a tool to oppose neoliberal globalization and as an instrument for the advancement of local socioeconomic development and the democratization of the agri-food sector. GI has been advocated as one of the strategies for the revitalization of agricultural communities and the production of quality food within the context of the neoliberal creation of global markets. In recent decades, the application of neoliberal policies has promoted enhanced global competition that has often created negative consequences for agri-food communities and actors while facilitating the concentration of political power and economic resources in the hands of a few large transnational corporations (TNCs). In this context, GI has been advocated as a tool for the reappropriation of culture and history in the definition of local agri-food production and practices and as a way to oppose the standardization and generalization of production practices and products associated with corporate agri-food (Allaire, Casabianca, and Thevenod-Mottet 2011; Bowen 2015; Parasecoli 2017). Moreover, GI has been described as an instrument to maintain fair and just competition and protect the well-being of small and medium-size agri-food producers and their communities (Bowen 2015; Sekine and Bonanno 2018; Vandecandelaere 2010).

The effectiveness of GI as an anti-corporate instrument should be also explored because of criticisms that denounce the process of the corporate appropriation of alternative spaces. According to these criticisms, GI has been transformed into a tool largely controlled by TNCs (Parasecoli 2017; Bonanno and Wolf 2018; Zukin 2008). Through the colonization of the state and emphasis on specific versions of the concepts of local and quality food, corporate forces advance discourses and practices that promote their economic interests while appealing to progressive ideals (Bonanno 2018; Sekine and Bonanno 2018). The colonization of progressive discourses is a particularly effective tool that allows TNCs to control life spaces that originally promoted anti-corporate postures (Bonanno and Wolf 2018). Additionally, as Parasecoli (2017, 17) notes, this process "indicates the pervasiveness of neoliberal theories." Following this scenario, the objective of this chapter is to review the theoretical assumptions and characteristics that define the now-dominant theory of neoliberalism and salient theories that advocate the regulation of markets and competition. The scientific relevance of this objective rests on the ambiguity with which neoliberalism and opposing socioeconomic theories are discussed in pertinent GI literature (Bowen 2015; Parasecoli 2017). Specifically, the regulation of markets and the corporate use of these measures that are identified as part of neoliberalism actually contradict basic tenets of neoliberal theory. Simultaneously, forms of market regulation are promoted as instruments that lead to economic growth. The contradictions associated with these opposing postures and the ambiguous use of neoliberal theory by TNCs constitute the core contributions of this chapter.

It opens with a brief review of the concept of GI and stresses salient historical events that defined its evolution. This section briefly reviews the basic characteristics of, and conditions that allow for, the development of GI-defining protocols such as the Paris Convention, the Lisbon Agreement and the TRIPS Agreement. The following sections illustrate the tenets of neoliberalism and opposing state intervention strategies. Neoliberalism is illustrated through the classic theories of F.A. Hayek and Milton Friedman and by their contemporary corporate adaptations. The latter refers to how neoliberal tenets are actually employed by corporate representatives. State intervention in the regulation of the economy is illustrated through instances of the Fordist model and its salient contradictions. The final section offers some comments on the limits of these theories vis-à-vis the phenomenon of GI.

Geographical indication and its evolution: a brief analysis

In pertinent scientific and political conversations, the contemporary definition of *GI* refers to the use of a name and/or symbol that is associated with and defines a specific agri-food product (Parasecoli 2017). This signifier/signified relationship finds its raison d'être in the desire to associate a particular agri-food product with a location, specific qualities, production procedures, reputation and the cultural milieu that define this product's existence (Bowen 2015; Echols 2008; Parasecoli 2017; Vandecandelaere 2010). GI stands at the intersection of the capitalism defining processes of *standardization of production/product* and *product differentiation* (Parasecoli 2017, 1). Standardization of production/product is part of that continuous tendency in capitalism to find the most efficient forms of production that reduce costs and increase quality and/or output. As prescribed by economic theory, desired forms of production generate competitive advantages as goods are produced at suitable costs and quality (Samuelson and Nordhaus 2009). Over the long run, however, these advantages tend to be eliminated as competitors adopt these production techniques and product characteristics. Accordingly, capitalist markets are characterized by processes of standardization of production as desired product quality, technology and labor use are incorporated into production.[1] The process of relinquishing rent positions (competitive advantages) is problematic socially and politically. Socially, it fosters conflict between those who favor and oppose the existence of competitive advantages. Those endowed with rent positions are generally unwilling to relinquish the benefits associated with competitive advantages. Those who do not enjoy a position of rent call for rent elimination, often in the name of fair competition. Politically, it constitutes a major point of contention among countries and regions because it favors the interests of some and penalizes the interests of others. Simultaneously, market competition promotes product differentiation. Product differentiation refers to the tendency of producers to achieve competitive advantages by enhancing one or more qualities of a product and, in so doing, making it distinguishable from and preferable to competing products. Product differentiation is achieved through a variety of strategies that include different techniques of production, prices, product quality and product availability/distribution. In essence, market competition entails contradictory demands associated with the simultaneous promotion of the standardization of production and product differentiation.

These conflicting demands are sources of discussion not only among scholars of capitalism and capitalist markets but, more importantly, among economic actors. In effect, the origins of GI rest on attempts to simultaneously promote and regulate competition. In the 19th century, as innovations and the expansion of markets characterized the growth of capitalism, there was almost no protection to the creation of new products, production processes and product differentiation. This situation was in part compatible with the then-dominant laissez-faire theory in that it contemplated the free circulation of goods and services (Smith 2009 [1716]). Simultaneously, however, classic liberalism advocated for fair competitions that involved the regulation of markets and the recognition of property rights, including those associated with intellectual property.

In this context, although patents existed in many countries, an international system of protection of intellectual property rights was lacking. In practical terms, this situation meant that patent applications had to be made simultaneously in a variety of countries, making the process of protection of intellectual property difficult to execute. It was precisely to address this situation that in 1878 an international conference was held in Paris with the goal of creating a system for the international protection of intellectual property. A draft of a pertinent proposal eventually resulted from the conference. This proposal was the subject of an international negotiation process, and an advance draft of the document was circulated for discussion at a second meeting in 1880. The final version of the proposal was eventually ratified in 1883 by a significant number of countries. This document is known as the Paris Convention: the oldest source of contemporary GI.

The expansion of agri-food markets and the significant growth of international trade that took place in the second half of the 19th century and first part of 20th century enhanced competition among agricultural producers. As competition increased, the demand for protecting local products against foreign competitors also increased, so governments implemented measures to safeguard local products. Despite the general acceptance of laissez-faire as the most appropriate economic theory, this policy was carried out through protectionist measures that were practiced by many countries and justified by the claim of the strategic importance of agriculture and food.[2] Additionally, the creation of barriers to free competition were also advocated to protect consumers from products of inferior quality (Echols 2008; Parasecoli 2017). It was understood that an efficient agri-food sector could specialize in low-cost commodities, such as those used as inputs for the production of industrial goods and basic foodstuff such as grains. But it could also produce high-quality products for the demand of quality consumers. In this context, the first organized attempt to protect local agri-food production took place in France – a country that to date retains the record for the highest number of GI-protected products – in the early 20th century (1905 and 1919), when the first laws on the protection of origins were passed. In Italy – the country with the second greatest number of GI agri-food products – similar measures were introduced much later, in 1962, whereas in Spain, earlier (1925) attempts to protect wine production in the La Rioja regions were formalized into law in 1970.

Increased efforts to use appellation of origin programs eventually prompted the signing of a more specific international agreement, in 1958. The Lisbon Agreement, as this new protocol was named, created a system whereby appellations of origins established within member countries were honored internationally. It further established the International Register of Appellations of Origins, which consisted of a list of all the products protected by the Lisbon Agreement. Various revisions of the original 1958 document were carried out through the years (1966, 1967, 1979), and by the mid 2010s, the Register included twenty-eight countries and more than a thousand appellations. In 2015, the Agreement was further emended to formally include the protection of GI and allow intergovernmental organizations to be parties in the Agreement. The most recent agreement concerning GI, however, is the WTO Agreement on Trade-Related Aspects of Intellectual Property Rights, or TRIPS. Signed in 1994, this document establishes basic standards for the protection and implementation of GI in signatory countries. In particular, it establishes procedures for transparency and accountability in the implementation of GI and guarantees measures against unfair competition. Despite these accomplishments, GI remains contested international terrain characterized by the clashes of two different visions.

These opposing views rest not only on contrasting interpretations of the role of GI but also on different readings of its strategic value as an economic policy. As far as the contrasting views of the role of GI are concerned, the European (EU) approach to GI is based on the established concept of terroir. Terroir refers to the environmental, cultural and historical properties associated with a geographical area. It follows that according to the European interpretation of GI, agricultural products and foods that use GI can be produced exclusively within the spatial and cultural contexts in which they originated and developed. Conversely, the opposing US-based understanding of GI rests of the notion of trademark. In this case, GI is not associated with a place, its culture and history. It is rather the private property of an owner (i.e., a company) that can produce that GI product anywhere it desires. In the case of GI as a form of economic policy, led by the European Union, some countries support the claim that GI is a tool that aids local rural economies, supports small and medium-size farms, safeguards the environment and promotes traditional food, local culture and authenticity in food production and preparation. Additionally, this view sees GI as part of processes that promote alternative forms of agri-food production and consumption (Bowen 2015; Vandecandelaere 2010). Represented by the position of the United States, the opposing view stresses that a system of pure free trade competition is more desirable. Accordingly, the protectionist nature of GI creates more problems than solutions by distorting the free functioning of the market.

Neoliberal theory, free markets, competition and the role of the state: two contrasting views

Neoliberalism has received significant attention since the late 1970s as the crisis of the Keynesian/Fordist social system required the implementation of new forms of regulation of the economy and society (Bonanno 2017; Harvey 2005; Mirowski and Plehwe 2009; Streeck 2016). Neoliberalism was originally formulated in the 1930s and 1940s as a new and different form of classic liberalism. Its creation was intended as a response to the emergence not only of totalitarian regimes – such as fascism and Stalinism – but also of democratic forms of governance that involved state intervention in the economy and society (Stedman Jones 2012). In this section, the basic tenets of classic neoliberalism are illustrated through a brief review of the works of its founding fathers: F.A. Hayek (2011 [1960], 1980 [1948], 1972 [1944]) and Milton Friedman (1982 [1962]; Friedman and Jacobson Schwartz 1971 [1963]). Their view of the economy contrasts not only classic liberalism ((Smith 2009 [1716]) and neo-classic liberalism (Marshall 1997 [1890]) but also policies advocated by TNCs in agri-food and other sectors. Accordingly, neoliberalism consists of two versions: one strictly academic and articulated in scientific publications and another political and implemented in legislative and regulatory measures.

The free market and competition

For classic neoliberals, there is an intrinsic relationship between political freedom and the free market in that the former cannot exist without the latter (Friedman 1982 [1962], 10–21; Hayek 2011 [1960], 169–171, 1972 [1944], 56–71). They contend that the free market guarantees political freedom because it eliminates the power of interest groups and the perils of the rule of the majority. For both Friedman and Hayek, in societies where the state intervenes in the economy, powerful economic groups employ lobbying, electoral contributions and similar strategies to affect parliamentary votes and majority-based decisions. Because these corporate interests often do not match those of the entire society, market distortions and social problems arise. Moreover, majority decisions always require compromise. Accordingly, they do not match all parties' actual objectives, resulting in incomplete, ineffective and dissatisfactory outcomes. Conversely, in a truly free market system, outcomes are the result of the impartial movements of supply and demand and are, therefore, always better suited than one-sided positions. In a free market and through voluntary participation, all parties agree to participate and knowingly accept the outcomes generated by this participation. This occurs because market transactions are always bilateral and because actors are confident that they have all the necessary information to make informed decisions. Also, if the exchange is not convenient, either party can freely withdraw. Therefore, the most important requirement in society is to maintain a free market system and prevent any form of coercion from altering this situation. Ultimately, the free market protects consumers, workers, producers and sellers because it allows them to freely seek alternatives. Critics call this view the economization of politics, whereby what is good for the market becomes automatically desirable for the entire society (Bonanno 2017; Brown 2015).

This classic view departs from the historical behavior of TNCs under neoliberalism. Also advocating free market principles, corporations operate to affect political decisions but also market outcomes. For instance, since the early 1980s, in a great number of countries – including the United States, Japan and the European Union – welfare programs have been either significantly reduced or eliminated (Harvey 2005; Streeck 2016). During the same period, however, state support of corporations and their economic activities has increased (Baker 2006; Bennett 2015; Kotz 2015). Known as corporate welfare, the expenditures associated with this phenomenon greatly exceeded state payments for welfare programs (Baker 2006; Bennett 2015). Simultaneously, support for free market reforms has not diminish the pressure that TNCs exercise on governments through lobbying and related strategies. According to specialized statistics, the number of accredited lobbyists in the United States has remained almost constant since the beginning of the new century, while lobby spending has drastically increased from US$1.45 billion in 1998 to US$3.15 billion in 2016 (Lobbying Spending in the U.S. 2016).

For Hayek and Friedman, free competition is the most effective way to allocate human, economic and natural resources. Departing from classic liberal theory, neoliberals contend that the fairness of competition should not be contaminated by political deliberations. In this respect, they would reject any argument contending that state regulation of markets – such as in the case of GI – would engender just and equitable forms of competition. They maintain that actual "fair" competition is open competition. This contention rests on their understanding of the concept of socioeconomic equality. For neoliberals, inequality, rather than equality, is a natural occurrence, ethically just, beneficial to society and the normal outcome of market competition (Friedman 1982 [1962], 161–163 (Hayek 2011 [1960], 164–165; Hayek1980 [1948], 30–32, 1972 [1944], 106–110). People are diverse, they stress. This diversity exists in nature, and at the beginning of competition, individuals are endowed with different skills and talents. It also exists at the end of competition as the use of these skills and talents discriminate among competitors, allowing some to be winners and others to be losers (Friedman 1982 [1962], 172). Like in the case of price formation, where the market establishes the appropriate value of goods and services, the appropriate value of individuals is determined through open competition (Hayek1980 [1948], 30–32, 1972 [1944], 106–110). Following 1940s functionalist analyses of social stratification (Davis and Moore 1945), classic neoliberals contend that some people prefer jobs that are easy to execute and involve relatively limited education, skills and risks. These are "pleasant" jobs, and this quality is discounted in the social and economic rewards that these individuals receive. There are conversely "dirty" jobs, which require higher levels of education and skills and involve an elevated quantity of risks and responsibility. These are jobs that are justly rewarded more and that ultimately constitute one of the primary sources of innovation and progress in society (Friedman 1982 [1962], 162–164).

Remaining silent on the objections that competition and the existence of skills and talents are socially constructed and based on the mobilization of accumulated resources that privileges the upper class (Davies 2014; Harvey 2014; Kotz 2015), Friedman and Hayek contend that the assumed negative consequences of high levels of inequality would be addressed through the expansion of the economy triggered by the free functioning of the market. Economic growth would eradicate poverty and other undesirable economic conditions, rendering state intervention unnecessary (Friedman 1982 [1962], 167–168). Moreover, free market competition allows for the matching of individual performances to required economic tasks in ways that are impersonal and commensurate to actual contributions to the economy (Hayek1980 [1948], 30–32, 1972 [1944], 106–110). This solution, they claim, avoids the undesirable situation in which performances and rewards are allocated following preestablished and politically motivated notions that are ultimately generated by arbitrary criteria and, more importantly, the exercise of power (Friedman 1982 [1962], 167). This efficient allocation of resources further justifies the superiority of the free market system over competing systems, particularly those based on the ideas of Marx and Keynes. For neoliberals, state-guided economies and the logic of politically established justice and fairness are contradictory because they could function only if the market logic is accepted. Given the fact that this is an impossibility, these economic theories are inconsistent and ultimately wrong (Friedman 1982 [1962], 167, 168).

Despite corporate pronouncements in support of open competition, the corporate understanding of competition differs from that of classic neoliberals. This difference can be illustrated through, what is arguably one of the most eloquent examples of corporate disregard for open competition: the 2007–2008 financial crisis. Through the two decades before the crisis, TNCs called for decreased state regulation of the economy, advocating the same neoliberal reasoning of the aforementioned advantages of free competition: The free market is the most efficient form of allocating resources (Wolf 2015). However, as the consequences of the crisis affected not only corporate profit but also the very existence of many of these companies, TNCs called for state intervention. Arguing that they were "too big to fail," TNCs stressed that the social consequences of their crisis would be highly detrimental to society in general and the working class in particular (Mirowski 2014). Claiming exceptionality, TNCs contended that it was the duty of the state to bail them out of their financial problems and restore their profitability. In an unprecedented move, the Obama administration in the United States de facto nationalized banks, financial institutions and companies in the productive sector. Once these companies' financial problems were addressed through the massive influx of public funds (corporate welfare), their leadership was reverted back to the very CEOs that engendered the crisis (Bonanno 2017). As stressed by pertinent literature, according to TNCs, the free market ought to be "free" as long as it is beneficial to them (Wolf 2015; Mirowski 2014; Streeck 2016).

Another relevant instance is the corporate support of intellectual property rights. For classic neoliberals, licenses and patents should not be regulated. Friedman (1982, 1962, 137–160) explains that the market should decide the desirability of a particular performance. Therefore, the market demand should regulate the production of commodities and the provision of services. If a commodity or a service is not desirable, he argues, it will not be in demand and vice versa. Accordingly, the physician who is not capable of effectively attending their patients will soon be without patients regarding of whether they have a license to practice medicine. However, in the case of corporate behavior, the existence of intellectual property rights involves not only the design to include industrial products but also natural resources such as plants, animals and biological products employed in agricultural and food production. Supporting TRIPS, corporations took advantage of the agreement and appropriated biological resources that historically were shared by local communities (Ritchie Dawkins and Vallianatos 1996).

The role of the state

Neoliberals' most decisive objection to forms of control of the market is expressed in their critique of the role of the state. Neoliberals disregard differences between Keynesian models and socialist proposals of market control to advocate a generalized objection to "state planning." Their position consists of three points. The first is the claim that the state regulation of the economy is coercive of individuals. The second refers to the distortions associated with the state manipulation of prices and restrictions on free competition. The third involves the question of the competence of state planners. As far as the issue of the coercion represented by state regulation is concerned, neoliberals contend that state-imposed planning denies the ability of people to act freely. It not only deprives individuals of their freedom to decide how to act but also creates a sense of dependency that curtails initiative and ingenuity. Coercion, they argue, should be rejected in favor of voluntary association. The state, therefore, should establish only the general principles that govern society rather than specific goals and behaviors that people must follow. In the case of GI products, their protection prevents producers and consumers from exercising their free will because they are forced to follow state rules rather than market trends and free competition. Moreover, these rules protect special interests and are not necessarily beneficial to society as a whole.

In the case of the state manipulation of prices and restrictions on free competition, Friedman and Hayek argue that the free market is the correct system for price formation. Prices set up through the free mobility of supply and demand are not only fair but also represent the exact value of commodities. In the case of agri-food products, their critique uses price support programs as an example of the many distortions that the state intervention generates. Often created to support small and medium-size family farmers, price support programs achieve opposite results because they advantage large producers that receive the vast majority of the available funds. Additionally, these programs tend to increase the prices of commodities to the detriment of consumers – in particular, lower class and poor consumers – and prevent investments in more productive areas/sectors. Accordingly, to support a relatively small number of farmers, a much greater number of consumers and producers are penalized. Finally, these programs promote inefficiency and the ineffective allocation of resources. This reasoning applies also to the case of GI. The impairment of the free functioning of the market and competition associated with the establishment of GI products creates inaccurate prices that cannot be associated with actual values. The inflation of prices promotes elite consumption that ultimately benefits the wealthy and larger producers (Friedman 1982 [1962], 181–182).

As far as the question of the competence of state planners is concerned, neoliberals stress that given the fact that the functioning of the market is based on the undisturbed concomitant movement of all pertinent variables, its outcomes are complete, impartial and fair (Hayek 1972 [1944], 48–49). This is not the case for state planning, though, because state officials are endowed with imperfect knowledge. This is always the case, even in situations in which the most sophisticated group of planners is involved. Despite their abilities, planners will never be able to accurately predict economic and social outcomes. Accordingly, their predictions and consequent plans will always depart from actual market outcomes. Moreover, any attempt that state planning could have to overcome the historical inefficiency of bureaucracy would not work, making this option highly undesirable (Hayek 2011 [1960], 166–183).

The Keynesian Fordist system

Based on Keynesian economics, Fordism was the system that dominated socioeconomic policies worldwide form the mid 1940s until the end of the 1970s. Despite its official end, some of its features are still employed in many countries, including EU member nations, emerging economies such as Brazil and India and developed nations such as Japan. The primary economic tenet of Fordism consists of the principle that the equilibrium between supply and demand cannot be reached through the autonomous functioning of the market. Additionally, Fordism centers on the contention that crisis recovery cycles engender harsh socioeconomic consequences that need to be addressed through state intervention (Aglietta 1979; Harvey 1990). While in classic Keynesianism, this is a condition that applies only to situations of crisis (Keynes 2009 [1936]), for Fordism, this is a constant condition that requires an equally constant management of the economy by the state. Accordingly, the state is the institution that is responsible for the well-being of the economy and society and, because of this responsibility, must legitimize its actions to the entire society (Habermas 1975). During the first four decades after World War II, Fordism worked well and permitted the concomitant expansion of the economy, the growth of salaries and wages and the expansion of profit. Simultaneously, it permitted the development of a large welfare state that reduced socioeconomic uncertainty and stabilized the lives of significant segments of the lower and middle classes.

One of the key conditions of Fordism was the nation-centered economy that dominated the period. This situation refers to the primacy of domestic companies and, in the case of multinational corporations, the existence of explicit connections between companies and nations. Under Fordism, the state assisted home multinational corporations politically, economically and militarily. In return, multinationals repatriated profits and invested locally. The underlying management labor pact created a condition whereby economic and political stability was sought in a deliberate attempt to avoid the establishment of extremist regimes of the right (fascism) and of the left (communism). Fearful of these particular occurrences and of the strength of the labor movement, corporations agreed to share power with labor and promote the – albeit partial – social inclusion of minorities.

In this context, markets were regulated and directed by states. The development of early forms of GI fit the Fordist scenario as desirable economic and social objectives were promoted through market control. Despite the almost four decades of success, Fordism and its regulation of markets ran into at least three insurmountable problems. The first consists of the fiscal crisis of the state. Market control required the growth of expenses that put fiscal pression on the state. As the state required additional resources to support market regulation, corporate activities and programs for the lower and middle classes, taxation was resisted because it diminished profit and was perceived as negative by wider segments of the population. Facing popular resistance and economic limits to the state ability to spend, state actions to control markets became ineffective and limited in scope (O'Connor 1984, 1973). The second limit refers to the inability of the state to concomitantly promote capital accumulation and social legitimation (Habermas 1975). Under Fordism, the regulation of markets and the satisfaction of the demands of subordinate groups limited the ability of the state to support corporate requests. Simultaneously, attempts to support the actions of corporations were seen as undesirable by the working and middle classes. Accordingly, the state was caught in a system based on conflicting demands that could not be reconciled. This inability to mediate demands for capital accumulation and social legitimation was the fundamental condition for the crisis of Fordism (Harvey 1990). Third, the nation-state cannot control transnational actors and global flows of resources. As documented by existing literature (e.g., Bonanno 2017; Robinson 2014), the bypassing of nation-state laws and rules has been one of the major outcomes of globalization.

Conclusions

The chapter offers a theoretical background that informs the discussion on GI. In particular, it elucidates the characteristics and limits of theories that frame the debate on GI. Classic neoliberalism is incompatible with calls for the development of GI in that it denounces the ineffectiveness of all forms of regulating the market. Analyses that see some episodes of the evolution of GI as a type of neoliberalism suffer from a limited understanding of the characteristics of this theory (Bowen 2015, 2010[3]). To be sure, the corporate version of neoliberalism contemplates state intervention to support corporate designs (corporate welfare) and forms of deregulation that are specifically advantageous for corporations. Even in this context, though, GI remains foreign to the understanding and applications of corporate neoliberalism. The compatibility of GI with revised forms of Fordism and its underlying Keynesian requirements is supported by the Fordist tenets that market ought to be regulated and that socioeconomic objectives should be established at the political level. It remains unclear, however, how neo-Keynesian policies address the historical limits of Fordism. In effect, the danger of these policies is multifaceted. They tend to reproduce conflict among regions and countries over fair trade. As these disputes continue, they tend to strengthen objections that stress the desirability and the assumed impartiality of the functioning of the free market. Additionally, these policies do not offer solutions to the fiscal crisis of the state, because state intervention is required to ensure the control of markets. Moreover, protectionist policies require state intervention to guarantee their justification (legitimation) vis-à-vis a host of actors that are disfavored by their implementation. In this context, the balancing of the processes of capital accumulation and social legitimation remains an unresolved question. Finally, the nation-state remains limited in its ability to control transnational actors and flows of resources. In light of these persisting conditions, the question of the ability of GI to be an anti-global neoliberal agri-food regime tool remains open. GI provides advantages and protection to some producers in a host of geographical areas and represents an incentive to the generation of quality food. Simultaneously, it does not transcend market relations and remains ultimately open to corporate control.

Notes

1 Standardization is also fostered by the enhancement of processes of exchange that demand compatibility among the various facets of production and distribution processes.

2 Among the advanced countries, only Great Britain adopted an economic policy that followed laissez-faire (Mathias 2001).

3 The otherwise excellent work of Sarah Bowen is an example of this fallacy, whereby neoliberalism is equated with capitalism. While neoliberalism is the currently dominant form of capitalism, it is not the same as the overall capitalist system. Accordingly, equating the two glosses over fundamental differences among the various forms of capitalism. This error problematizes the development of a clear understanding of the socioeconomic implications associated with different capitalist regimes and the formulation of effective alternatives.

Bibliography

Aglietta, Michel. 1979. *A Theory of Capitalist Regulation*. London: New Left Books.

Allaire, Gilles, François Casabianca, and Erick Thévenod-Mottet. 2011. "Geographical Origin: A Complex Feature in Agri-Food Products." In *Labels of Origin for Foods*, edited by Elizabeth Barham and Bertil Sylvander, 1–12. Wallingford: CAB International.

Baker, Dean. 2006. *The Conservative Nanny State: How the Wealthy Use the Government to Stay Rich and Get Richer*. Washington, DC: Center for Economic and Policy Research.

Bennett, James. 2015. *Corporate Welfare: Crony Capitalism That Enriches the Rich*. London and New York: Routledge.

Bonanno, Alessandro. 2018. "Best Practices: The Artificial Negativity of Agri-Food." In *Resistance to the Neoliberal Agri-Food Regime: A Critical Analysis*, edited by Alessandro Bonanno and Steven A. Wolf, 21–34. New York: Routledge.

———. 2017. *The Legitimation Crisis of Neoliberalism*. New York: Palgrave Macmillan.

Bonanno, Alessandro, and Steven A. Wolf. 2018. *Resistance to the Neoliberal Agri-Food Regime: A Critical Analysis*. New York: Routledge.

Bowen, Sarah. 2015. *Divided Spirits: Tequila, Mezcal and the Politics of Production*. Oakland, CA: University of California Press.

———. 2010. "Embedding Local Places in Global Spaces: Geographical Indications as a Territorial Development Strategy." *Rural Sociology* 75 (2): 209–43.

Brown, Wendy. 2015. *Undoing the Demos. Neoliberalism's Stealth Revolution*. New York: Zone Books.

Davies, William. 2014. *The Limits of Neoliberalism. Authority, Sovereignty and the Logic of Competition*. Los Angeles: Sage.

Davis, Kinsley, and Wilbert Moore. 1945. "Some Principles of Stratification." *American Sociological Review* 24: 757–72.

Echols, Marsha A. 2008. *Geographical Indications for Food Products. International Legal and Regulatory Perspectives*. Amsterdam, NL: Kluwer.

Friedman, Milton. 1982 [1962]. *Capitalism and Freedom*. Chicago: University of Chicago Press.

Friedman, Milton, and Anna Jacobson Schwartz. 1971 [1963]. *A Monetary History of the United States, 1867–1960*. Princeton, NJ: Princeton University Press.

Habermas, Jürgen. 1975. *Legitimation Crisis*. Boston: Beacon Press.

Harvey, David. 2014. *Seventeen Contradictions and the end of Capitalism*. Oxford: Oxford University Press.

———. 2005. *A Brief History of Neoliberalism*. Oxford. Oxford University Press.

———. 1990. *The Condition of Postmodernity*. Oxford: Blackwell.

Hayek, F. A. 2011 [1960]. *The Constitution of Liberty*. Chicago: University of Chicago Press.

———. 1980 [1948]. *Individualism and Economic Order*. Chicago: University of Chicago Press.

———. 1972 [1944]. *The Road to Serfdom*. Chicago: The University of Chicago Press.
Keynes, John Maynard. 2009 [1936]. *The General Theory of Employment, Interest and Money*. New York: Classic Books America.
Kotz, David. 2015. *The Rise and Fall of Neoliberal Capitalism*. Cambridge, MA: Harvard University Press.
Lobbying Spending in the U.S. 2016. "Online Document." Accessed March 2, 2017. www.statista.com/statistics/257340/number-of-lobbyists-in-the-us/.
Marshall, Alfred. 1997 [1890]. *Principles of Economics*. New York: Prometheus Books.
Mathias, Peter. 2001. *The First Industrial Nation: The Economic History of Britain 1700–1914*. London: Routledge.
Mirowski, Philip. 2014. *Never Let a Serious Crisis Go to Waste*. London: Verso.
Mirowski, Philip, and Dieter Plehwe, eds. 2009. *The Road from Mont Pèlerin: The Making of the Neoliberal Thought Collective*. Cambridge, MA: Harvard University Press.
O'Connor, James. 1984. *Accumulation Crisis*. New York: Blackwell.
———. 1973. *The Fiscal Crisis of the State*. New York: St. Martin's Press.
Parasecoli, Fabio. 2017. *Knowing Where it Comes from: Labeling Traditional Food to Compete in a Global Market*. Ames, AI: Iowa State University Press.
Ritchie, Mark, Kristin Dawkins, and Mark Vallianatos. 1996. "Intellectual Property Rights and Biodiversity: The Industrialization of Natural Resources and Traditional Knowledge." *Journal of Civil Rights and Economic Development* 11 (2): 431–53.
Robinson, William, I. 2014. *Global Capitalism and the Crisis of Humanity*. New York: Cambridge University Press.
Samuelson, Paul A., and William D. Nordhaus. 2009. *Economics*. New York: McGraw-Hill.
Sekine, Kae, and Alessandro Bonanno. 2018. "Geographical Indication and Resistance in Global Agri-Food: The Case of Miso in Japan." In *Resistance to the Neoliberal Agri-Food Regime: A Critical Analysis*, edited by Alessandro Bonanno and Steven A. Wolf, 106–19. New York: Routledge.
Smith, Adam. 2009 [1776]. *An Inquiry into the Nature and Causes of the Wealth of Nations*. Blacksburg, VA: Thrifty Books.
Stedman Jones, Daniel. 2012. *Masters of the Universe*. Princeton, NJ: Princeton University Press.
Streeck, Wolfgang. 2016. *How Will Capitalism End*? London: Verso.
Vandecandelaere, Emily. 2010. "Geographic Origin and Identification Labels: Associating Food Quality with Location." In *Innovation in Food Labeling*, edited by J. Albert, 137–52. Rome: FAO.
Wolf, Martin. 2015. *The Shift and the Shock: What We've Learned and Have Still to Learn from the Financial Crisis*. New York: Penguin Press.
Zukin, Sharon. 2008. "Consuming Authenticity: From Outpost of Difference to Means of Exclusion." *Cultural Studies* 22 (5): 724–48.

Part II

The Asian context

2 Geographical indications out of context and in vogue

The awkward embrace of European heritage agricultural protections in Asia

Hart N. Feuer

Introduction

Through the presently hurried phase of policy deployment in Asia, geographical indications (GIs) for agri-foods have risen feverishly to the level of a public discourse while quietly expanding the room to maneuver vis-à-vis the European GI system and global trade rules. This state of affairs has emerged due to the confluence of (1) the enthusiastic but overly simplified lay understandings of GI policies driving political motivation, (2) the eager support and lenience of European engagement and (3) the nascent diversity of models on which new GI laws can draw. This position is a corollary of that of Bonanno (this volume), who highlights the conundrum that GI policies are expected to achieve seemingly contradictory goals, such as instituting preferential domestic intellectual property rights while promoting free trade or privileging restrictive production specifications while advocating for market expansion. The fact that European foreign policy would not only tolerate but actively perpetuate such contradictions suggests that the room for experimentation within and between the neoliberal free market approach and the state-interventionist approach is wide. In other words, the European approach to promoting GI intends not to consistently replicate European norms of sui generis GI in the upcoming markets of Asia but rather to shore up their position vis-à-vis the United States by drawing new countries into its general orbit of intellectual property rights for agri-food. Through a comparison of the approaches in Japan and Cambodia, which accentuates the different policy approaches and availability of public resources, I show how the awkward embrace of GI policies in Asia reflects both the underlying ambiguities of "place of origin" policy recruitment and the incongruities faced in, and created by, different national contexts.

The devious charisma of the geographical indication approach

An observation not lost on many academic and civil society commentators is that GI is able to generate a level of enthusiasm and social mobilization that is unusual in the otherwise dry field of intellectual property rights. Despite being a cousin concept to more run-of-the-mill rights mechanisms, such as trademarks, copyrights and patents, GI is not a commonly accepted fixture of a well-established national legal system. Rather, GI policies are frequently "rediscovered" in new countries and regions in a process that often sparks renewed excitement and debate (Calboli 2017, 4; Heath 2017; Hughes 2017, 65–67). This characteristic of GI policy has garnered it the title of the "Sleeping Beauty of the intellectual property world," a euphemism coined in 1995 (by Florent Gevers during a symposium of the World Intellectual Property Organization (WIPO)) and trumpeted again in 2007 (by Marcus Höpperger at another WIPO symposium). This confidence is a natural response to the innocent notion that the "basic concept underlying GIs is delightfully simple" (March 2007, 8, from the official magazine of WIPO). And indeed, while the vagaries of codifying national copyright and patent laws rarely raise an eyebrow among regular people, GI policies often capture public imagination because they seem to correspond with existing reality on the ground. A glance at public speeches and media portrayals of newly minted GI laws in East Asia reveals variations of the following lay perception: *Now here is a legal protection that normal people who care about our national food can appreciate* (e.g., de Gaudemar 2016; Passeri and Chheng 2016; UNCTAD 2015). At work behind such judgments is the expectation of achieving some degree of parity with the product differentiation and competitive advantage established by Europeans through historical GI policies, based on the naïve normative assumption that intrinsic qualities of food products can and should be readily recognized worldwide. As many promoters have suggested over the decades, if the quality, know-how and connection to geographical origin are historically present, they must "simply" be valorized (Addor, Thumm, and Grazioli 2003). So, if patents and copyrights reside in the rarefied space for valorizing high technology innovations, GIs appear to realize the obvious: confirm what is already widely known and protect it against fraud and quality degradation.

And yet, the type of the sui generis model of GI currently spreading across Asia is often reductive and technical, which can make it inadequate for capturing what appears to be "widely known" as this knowledge arises from a fluid sense of authenticity and quality. A comparable oversimplification surfaced with expectations about organic agriculture in developing countries, again with proponents assuming that certification would readily valorize the existing quality and know-how of the many "default organic" farmers worldwide, when in fact such farmers were often simply resource-poor rather than agronomically adept (De Launey 2005; Källander and Rundgren 2008, 29; UNESCAP 2002). In this vein, there are inflated hopes pervading across Asia that GI policies will draw the attention of the world to the manifest achievements of each nation's agri-food system. To be fair, policymakers and most experts are more restrained and nuanced in their push for GI, at least on paper. However, I argue that the public enthusiasm for a renewal of traditional quality and care, coupled with the increasing discursive power of GI, has created an inexorable pressure to pursue GI policies regardless of their strategic suitability. The constellation of promoters of sui generis models of GI – including oriGIn, SINER-GI, WIPO and the French agency for development assistance[1] – typically abstain from sweeping statements endorsing GI policies for each country, but in the end, they are often persuasive. Typically, the pitch begins by noting that GI is already inscribed in TRIPS and other multilateral agreements signed by most countries, so there is no reason not to use this existing opportunity; all that remains is to evaluate one by one the feasibility of different GI-candidate products. This seemingly innocuous expression of contingency preempts more systemic critical response while opening the door to the mission creep inherent to project-based action. At this stage, governments and the public quickly discover or receive a list of "potential GIs," on which are found a nonrepresentative collection of readily applicable agri-food products (the low-hanging fruit). This serves the purpose of communicating to a proud public that a sui generis GI policy is evidently suitable in their country. What appears at this point as "delightfully simple" is a prelude to a complex GI process that guides and channels how "things," which are in this case fluid mental frameworks and rumors about quality, become commodities.

The process of engendering a sense of inevitability imbues GI policies with the expectation that they can independently create traction against the pull of commodification. In reality, the *direct impacts* of GI policies touch on only one pole ("deepening") of the comprehensive engagement required for integrated rural and agricultural development, which has been codified so well in van der Ploeg and Roep's (2003) triangle of differentiation (deepening, broadening, regrounding). Rather, it is the indirect impacts which are generated through the GI policy mechanism pressures or the very nature of the GI institutionalization that serve to bring together all three poles of rural development. Outwardly, the purpose of GI as certification is to instill in (increasingly distant, unconnected) consumers the credible belief that GIs are "deconstructed commodities" with differentiated value as opposed to trivialized, homogeneous "commodity goods" (Augustin-Jean 2012b, 57). Whether the results in each case of a GI product provide sufficient deconstruction of a commodity to increase its value depends on the details and scope of the valorization effort.

While the notion that the broader product valorization strategies realize "secondary impacts" (on biodiversity, traditional knowledge, community-building, etc.) that should be documented and valued alongside the usual metrics of higher prices, sales and employment (Bramley, Biénabe and Kirsten 2009, 118), analysts tend to use relatively simplified, individual GI products as the units of analysis in performance measurement. This reflects the urgency of the case study approach for helping to evaluate, optimize, or troubleshoot GI policies and assess rents accrued to participants; doing so, however, often leaves the broader national movements of the GI policy relatively under-analyzed. This chapter will draw from examples of GIs (and their discontents) in Japan and Cambodia, but it will do so following a perspective that views GI more discursively – as an evolving expression of the power of policy proponents and the agency (in the actor-network sense) of the GI products themselves. This view aligns with that of van

Ittersum et al. (2003), who understood the "buzz" around GI, particularly at the domestic level, as an end itself rather than as merely the means for enhancing the impact of the individual GIs or realizing high-value exports. More simply, this orientation can be summarized by an observation of a key informant in Cambodia, who writes in a United Nations (UN) Food and Agricultural Organization (FAO) report that Cambodian authorities and non-governmental organizations (NGOs) have recently been encouraged by international organizations to establish a "GI culture" (Prak 2010, 39).

Innocent embrace of GI policy – critical stance toward the rules of the game

As with many interventions imported in the postcolonial era that channel development through precolonial institutions, developing countries have learned strategies for appropriating the renewed financial flows and logistical support attached to new trends. Especially when the intervention in question is contested by the world powers, as the sui generis GI vs. trademark models of agri-food are (Blakeney 2015, xiii), latecomer countries being courted by policy diplomats can proactively survey the field and elect approaches that are radically different from the vision of core promoters. In this case, both Japan and Cambodia have embarked proactively on the development of GI policy, but with different institutional approaches.

The tactics of the Japanese Ministry of Agriculture, Forestry and Fisheries (hereinafter simply Ministry of Agriculture) for designing their sui generis GI law revealed their objective of optimizing the GI framework for Japan by privileging the views of countries that are less vested in, and presumably less burdened by, policy inertia developed a hundred years ago (on this latter point, see Raustiala and Munzer 2007). Among other documents released publicly, their policy drafting process included a 623-page working paper from 2011 commissioned by the Ministry of Agriculture that contains the raw output from surveys of third-country GI laws and interviews with relevant staff in different countries (IIP 2011). The surveys were conducted mostly with "GI lightweights" such as Germany and the United Kingdom but also "GI skeptics" such as China, South Korea and the United States (Calboli 2006). The resulting Japanese sui generis GI policy in 2015, which involved input from stakeholders from the food industry, academia, farmers' unions and foreign experts, can be understood either as a poorly consolidated mishmash of various national approaches or as a keen integration of competing policy goals (Kojo 2006). What it most clearly represents, however, is a deliberate departure from the European model that actually shores up the preexisting Japanese regional collective trademark GI system in many ways but that is subtle enough not to viewed as a rebuke of the EU system from which Japan intends to benefit (Port 2014).

This is achieved by leveraging the European Union and the WTO by augmenting intellectual property enforcement for marquee export-oriented products and oft-defrauded products (e.g., matcha powder, top-grade green tea, dried kaki persimmons, wagyu beef, premium melons, etc.) while more quietly exploiting the rest of the legal opening provided by TRIPS to prop up more vulnerable, domestically important products/regions with marketing assistance and public sector legitimation. This is evident in both the intention and pattern of GI registrations since the GI law went into force in 2015. For one, and in contrast to the Cambodian case, the Japanese sui generis GI law is implemented by the Ministry of Agriculture rather than an intellectual property division,[2] which essentially adds GI to the portfolio of rural support mechanisms in Japan. This helps explain how, in seeming violation of the European spirit of GI laws, the Ministry of Agriculture expected to register an astounding number of 1,000 non-alcohol products (Kojo 2006, 14); for reference, Italy has registered 274 non-alcohol GIs, whereas the whole EU has just over 1,300 (according to the European Union's database of origin and registration (DOOR), as of October 2018). Although expectations have since been tempered about this number, the goal of channeling support to smaller production areas in Japan was proudly announced in an informational session for prospective GI candidates in the Nara region in 2015, the first year of the policy.[3] This goal has already partially come to fruition with the successful GI registration and subsequent marketing support (e.g., website, local events, media coverage) provided to relatively obscure and small-scale agricultural products such as national varieties of shellfish, eggplant, turnip and burdock root. In the meantime, owners of the more iconic brands/GIs have been emboldened to chase down their intellectual property rights in China, in Europe and domestically.[4]

Cambodia, under the Ministry of Commerce (Department of Intellectual Property, or DIP), has also exploited the recent worldwide GI turn but has taken a different strategic approach more in line with its developing country status – that is, by harvesting development aid and logistical support from European sources for expensive marquee GIs while incrementally cultivating local capacity and starting domestically important initiatives in the background. Central to this strategy has been the privileging of practical learning about the technical aspects of GI, which was largely achieved by allowing certain flagship GIs to begin institutionalizing themselves even before the Law on Geographical Indications was finalized in 2014. Beginning in 2007, the Cambodian authorities let the French development agency and FAO guide much of the process and help draft the law. As the head of the DIP noted in 2017, this allowed them to "see how the French would do it and learn how to make [their] own adjustments along the way."[5] The result was that the two initial GI registrations in 2010 for Kampot pepper and Kampong Speu palm sugar, which were entirely funded by international organizations (WIPO, FAO) and European aid (mostly French bilateral assistance), fell closely in line with EU norms; simultaneously, this experience provided Cambodian authorities a crash course in GI implementation years before the relevant policy was enacted. This achieved two parallel goals: maintaining an open lifeline to development aid for costly export-oriented GI valorization and building up local capacity for later attempts to economize on more domestically important GIs.[6]

Indicative of this bifurcation is the prospective gap between these initial GI registrations and most of the GIs in the pipeline. The first two products are produced or marketed in ways that make them largely irrelevant to the Cambodian domestic market while requiring significant investments in promotion and export. Kampot pepper has been transformed almost entirely into a souvenir and a prestige ingredient for foodies and European haute cuisine, while Kampong Speu palm sugar is mostly processed into granules, a product form which is alien to Khmer cooking.[7] Both products are also certified organic to European and other international standards, but most brands have ignored the domestic certification from the Cambodian Organic Agriculture Association (COrAA). In any case, the next products in the GI pipeline, Koh Trung Pomelo and Kampot sea salt, are less blatantly neocolonial: They are likely to be purchased as gifts or for ceremonies and may be exported as well.[8] Subsequent GI products, moreover, are far more likely to gain local traction because they are not widely known outside of Cambodia and Khmer cuisine or have properties (such as fishy smells or exclusive local culinary use) that do not lend themselves to wide (international) promotion. In this sense, the intended result is comparable to the Japanese case: heavy investments in valorization and international promotion for symbolically important and EU-approved products, followed up by more domestically relevant products and local promotion.

In Cambodia and Japan, concerted efforts have been made to appropriate the legal and logistical benefits, as well as the enthusiasm, generated by the successful expansion of European GI policy while also preparing the ground for (or reinforcing) more locally relevant and utilitarian product promotion. Demonstrating this dramatically, the Cambodian Minister of Commerce, H.E. Pan Sorasak, led a memorable delegation to Geneva in March 2018 to be the first country to deposit instruments of ratification to the Geneva Act of the Lisbon Agreement on Appellations of Origin and Geographical Indications. His remarks on the occasion, that Kampot pepper is proudly the first GI product in Southeast Asia to receive EU protection, outwardly communicate his wish to enable international protection for Cambodia's excellent products with national origin (Khan 2016), but he is also realistic about the international prospects for many of Cambodia's less evidently famous products. Rather, he is more likely aiming to use the GI law to protect Cambodian producers from competition from neighboring countries in Southeast Asia, which is why the GI law is implemented in the Department of Intellectual Property rather than the Ministry of Agriculture, Forestry and Fisheries. This concern is reversed in Japan, a country with declining agricultural production that is less concerned about fraudulent sale of little-known traditional products and more hopeful about creating a strong pretext for protectionist rural development measures. This inward orientation in both countries reflects, on the one hand, low expectations about promoting many unique traditional products abroad and, on the other hand, a need to promote domestic products locally or regionally in the face of declining consumer interest and skill. The former point is not unfamiliar to the European case, where many countries have registered obscure or small-scale GI products that are not destined for international (or even extra-regional) audiences. The latter point, which is more proactively recognized in Asia, is more of a latent realization in the EU countries that GI certifications will increasingly fall on deaf ears if consumer food knowledge (and willingness to pay) does not keep apace (Chabrol and Muchnik 2011); in other words, a continuous process of revalorization is necessary for each product (category), or more integrated processes of rural development similar to van der Ploeg and Roep (2003) are necessary. Asian countries with fresh GI laws seem to be aware of the longer-term problems of the sui generis system and are preemptively moving to avert them by crafting more locally adapted policies while outwardly still playing by the old rules. This suggests that the cultivation of a "GI culture," as opposed to simply the implementation of GI policies, is a more strategic endeavor meant to fully use the support structures associated with the "GI turn" in recent decades.

For geographical indications, failure is often success, or at least a surprisingly productive endeavor

The "GI turn" has created inertia in the directions of product valorization, certification and international trade such that turning away from the constructed sense of inevitability of GI can take considerable self-reflection. Even as a researcher in the field, I was swept up in this fervor to such an extent that I found myself feeling uncritically disappointed upon learning that various producers had given up on their GI registrations or were left out of a promising GI initiative. Later, I came to discover, or was convinced by other observers, that these events were for the best or were at least strategic in nature. Indeed, often the "prospect" of a new GI, as opposed to its eventual realization, is sufficient to engender change by galvanizing various secondary changes. And even GIs with direct outcomes that are considered poor or undesirable can have notably positive impacts if the scope of evaluation is adjusted or reframed. Consider, for example, the distraught Normans when, in 1926, an appellate court in France ruled that camembert was henceforth a generic term. This loss of monopoly power, however, quickly led to even greater appreciation for camembert from Normandy and a worldwide bonanza for camembert styles that has benefited the French (Boisard 2003, 16–18). In understanding how GI policies are being embraced and negotiated in Asia, it is important to view not only planned but also contingent processes of reflection and response. These are reviewed in the next section, with particular attention drawn to particular processes of realignment (of the domestic market, expectations, allies), spillover (into secondary markets) and resolution (of historic conflicts).

Failure to consolidate a definition of geographical indication

While GI laws often require applicants to prove that their products are historical – that is, they have been produced in a comparable manner for at least a certain period (twenty-five years in Japan and the European Union) – there is often no requirement for the oldest, strictest, most traditional or most natural characteristics to be codified. The "strawberries of Sologne" described by Augustin-Jean (2012b, 60–61) exemplify a case in which quality does not arise from exclusively localized tradition and know-how *from* Sologne but rather from a concerted effort *in* Sologne to standardize good practices from around France. In this sense, its connection to place is illusory at best, while its destruction of strawberry diversity and reductiveness toward strawberry farming systems stand largely in opposition to the socialized expectations of GI as the vanguard of local traditions. A comparable case is on display currently in Japan in the conflict over the definition of Hatcho miso, a red fermented bean paste that has continuously been produced in Hatcho district in Okazaki, a city in Aichi Prefecture, for almost 700 years. A GI for Hatcho miso was granted in December 2017 to a consortium of more recent red miso producers from other parts of Aichi Prefecture, with the oldest two producers in Hatcho district refusing to join on the grounds that the standards proposed by the other producers were not strict enough to be respectful of the product's storied history.[9] Subsequently, the historical producers from Hatcho district submitted a formal complaint to the Ministry of Agriculture and have initiated a petition to protest their treatment. This form of "competition," to impose certain standards and relativize authenticity, again disrupts popular expectations about a GI policy's capacity to neatly consolidate the proposed products. Instead, the GI registration offers a public platform and some urgency for the resolution of simmering internecine struggles, as well as some less binary methods (compared to trademarks) for recognizing subtle quality differences among heterogeneous products (Allaire 2012, 85). The potential to leverage this aspect of GI legislation, however, depends on the manner in which the narrative of the product is socially constructed, and this in turn is subject to power relations between producers and to the whims of government arbitration.

Failure to establish a seemingly promising geographical indication

The case of persimmon growers in the Nara region of Japan to opt out of a GI registration process represents an ambivalence toward an inscription process that imposes a rigid or reified production structure. This consortium of producers decided not to apply for a GI after hearing a promotional session on Japan's 2015 GI law and despite the enthusiastic support of a member of their local city council. Although they understood the "honor" associated with receiving a worldwide certification, the producers noted that they had already achieved fair marketing conditions on their own accord. They had a well-functioning producers' cooperative and had developed short value chains through their direct-order model, such that they internally maintained and controlled the quality of persimmons sold to their regular customers in Japan and China. The onerous GI registration and associated limitations on flexibility imposed by fixed production guidelines seemed an unnecessary burden for their cooperative. In fact, the realization that their operations were already ahead of the curve was a source of pride and motivation.[10]

In contrast, the case of a proposed GI for oranges in Cambodia reflects the challenges faced when trying to artificially impose a narrow frame of reference onto a product and production area with a high informal reputation but little preexisting technical or social consolidation. The "Pursat orange" of Cambodia, which is in fact green in skin color, was transformed into a sought-after product during the French protectorate, when colonial botanists worked to improve local varieties of oranges. Their research station, in Dap Bat commune of Pursat province, was so well-known that the colloquialism "Pursat orange" became commonplace. Nearby districts in Pursat province developed well-known orange plantations, but so did many districts in neighboring Battambang province.[11] Eventually, Battambang-sourced oranges became more famous due to their sweetness, leading to the awkward linguistic result "Pursat oranges of Battambang" (*kroich pousat battambang*). Although the citrus itself and many of the associated farm practices arose in Pursat province, this arose from colonial (i.e., foreign) imposition rather than the natural evolution of soil and know-how that is fetishized for geographically indicated products. In fact, the ensuing expansion of orange growing to nearby districts of Battambang province and the accompanying improvement in reputation is more indicative of this natural process of refinement, but in transgressing an administrative border, the orange's name began to be at odds with its highest quality production area. This has more recently led to a conflict between the producers in Battambang and Pursat, with each accusing each other of appropriating part of the product's identity: either its history or its currently high reputation. The imported concept of sui generis GI has, for example, emboldened many new settler farmers in Pursat with limited know-how and untested growing conditions to argue that their product inherently deserves the title of Pursat orange. Reconciling these simmering struggles is not a romantic consummation of a historic product, but it can, even when it fails, achieve other important secondary impacts.

As in the example of the strawberries of Sologne and wine from the Republic of Georgia (see Shtaltovna and Feuer, this volume), the Pursat orange lies at the confluence of efforts to consolidate and promote a unified product, whose differentiation and quality have emerged in a multipolar way. The consistent failure to reach an agreement about the origin and name was a drag on the Cambodian regional orange economy and left both provinces vulnerable to even more severe fraud from oranges imported from Vietnam. The GI negotiation process pushed these parties to important technical and social reflections. Battambang province orange growers with excellent quality are now working to more concretely specify the uniqueness of their product and have begun earnest negotiations with orange growers in Pursat province to amicably resolve questions of the naming and origin of these oranges, as well as to work together to fend off mislabeled imported oranges.[12]

Failure to achieve socially optimal goals of a geographical indication

Even in the most lauded cases of establishing new GIs, such as that of Kampot pepper in Cambodia, the particulars of the welfare impacts of GI are important to note. Such analyses are not uncommon upon the establishment of a GI (Bramley, Biénabe and Kirsten 2009, 122), but the distribution of rents invariably evolves over time and therefore should be re-evaluated periodically. And indeed, some of the sharpest criticism and overall pessimism toward GI is directed at this progression (Belletti, Marescotti and Touzard 2017; Kerr 2006). Despite the celebrated rollout of the Kampot pepper GI, and the encouraging farmgate and retail prices realized,[13] the welfare impacts have remained under scrutiny. The president of the Kampot Pepper Promotion Association (KPPA) noted that although he enjoys his new status and invitations to black-tie events, "the Kampot pepper GI is not really for Cambodians."[14] This was in reference to the three most common worries associated with developing country GI policies (Augustin-Jean 2012a, 7). The first, and perhaps most glaring, is that almost the entire GI pepper crop is purchased by foreigners (70 percent is exported, and most of the remainder is sold locally as souvenirs), thereby economically and structurally excluding most Cambodian consumers. This is associated with the second worry, namely that the bulk of the profits (including for secondary markets such as ecotourism) are absorbed by intermediaries, most of whom are external to the territory or foreigners.[15] Here, the French colonial history is particularly obvious, with much of the processing and packaging conducted by French nationals and exported to France, layered on the historical detail that Kampot pepper is a French colonial-era construct associated with the French "discovery" of excellent pepper in Cambodia and its export through the seaport of Kampot (Direction de l'Agriculture et du Commerce 1899; Kitagawa 2005; Société de géographie commerciale de Bordeaux 1900). The third worry, now being realized, is that corporate raiders or foreigners will encroach on the valuable land and process of pepper production itself. In 2017, Chinese and Singaporean firms bought up major plantation size tracts of Kampot pepper land, adding to the existing landscape of major American, French and Australian ownership (Cheng 2017). These issues have tarnished the GI's reputation for rural development.

From a minimalist perspective, however, the Kampot pepper GI should still be considered a success. Whether farm laborer, manager or distributor, there are higher salaries and more pepper-related jobs for Cambodians in Kampot than other cultivation areas. And yet, other cultivation areas, particularly those with comparable quality to Kampot, are cashing in on the notoriety of Cambodia for pepper and the retreat of the Kampot area from the domestic market. The historically excellent producers from Kirivong-Takeo province have gained relatively in domestic stature, while provinces like Koh Kong and Mondulkiri have produced excellent-quality products that have supported the domestic organic certification scheme. Fraudulent use of the Kampot pepper name in domestic markets is also decreasing due to the renewed interest in alternative production sites and absence of affordable certified Kampot pepper. In short, the illusory and fraudulent dominance of the Kampot namesake is being replaced by more openness and plurality in the domestic pepper market.

Conclusion: geographical indications – a box of unintended tools

It is tempting to believe that the emphasis on trade and value generation by agri-food GIs has been the primary driver behind the success of the European promotion of sui generis GI policies in Asia, but the two country cases in this chapter demonstrate that Asian governments are discovering more utility in subsidiary aspects of GI, such as differentiating domestic products, privileging national producers and slowing quality erosion in agri-food products. Producers are also discovering that, as in Europe, GIs provide an inconsistent bulwark against domestic corporatization and/or appropriation of value by foreign entities. In developing countries, such as Cambodia, the disregard for the domestic market and appropriation of locally created value by foreigners in the early GI products smacks of neocolonialism and can even be argued to be a trojan horse for aggressive neoliberal market forces to gain a foothold in these rural areas. In richer countries, like Japan, the advent of the sui generis GI law provided cover for the government's claim that they are protecting the interests of traditional agri-food producers, but their actions have continued to privilege corporate actors and protect export industries (Sekine and Bonanno 2017, 13–38). Asian governments' interest in GI thus arrives at a time when many are pushed to support national agricultural patrimony but are also pulled to maintain their participation in the project of world trade liberalization.

These situational conditions for the emergence of GI in Asia also align with a more fundamental association of food and territory in many culinary imaginaries. So, in part, the post-2000 "GI turn" can be explained by the aggressive promotion by the European Union of its preferred mechanism of intellectual property rights, but its speedy blossoming arises from the internal attractiveness of sui generis GI, particularly for Asian countries with a preexisting tendency to associate place with quality. The more fundamental and optimistic initiative to legitimate GI-like products is seen as separate from the otherwise pessimistic perception about the extent of capture of the world heritage food trade by European countries (including sweets, meats and alcohol, among others). Of course, it is not uncommon for strong, unique flavors such as Scotch whiskey or Roquefort cheese to gain traction in foreign milieux, but it is perhaps rational for many Asian countries to remain sanguine about the prospects for odorous or unusual products such as Cambodian fish paste (*prahok*) or Japanese fermented soybeans (*natto*) to make a splash in world markets. Perhaps this is why the trajectory of many Asian GI policies appears to be one of initially subscribing to the world GI community based on a few emblematic products while ultimately aiming to appropriate the GI concept for domestically relevant agri-food promotion. This potentiality is not, in fact, contrary to the European conception of international GI promotion (Augustin-Jean 2012a, 5), which expects endless bargaining and different outcomes related to each country's history, culture and economy – not to mention a cohort of GI products that is unlikely to become well-known outside of each country.

Part of the attraction to GI policies in Asia is the recognition that GIs can be used as an endogenous platform for demonstrating the value, and encouraging the continued cultivation, of consumer food skills that are necessary for society to reproduce interest in traditional cuisine. In many of the gift-giving cultures across East Asia, one is disproportionately honored for presenting agricultural products and being able to render their quality in narrative and technical terms (Lin and Mao 2015). Unlike many value-added certifications, GIs do not have a linear and easily understood impact on food purchases; they demand that consumers are already aware and motivated to buy certain regional specialties, and then they provide a guarantee of origin and consistency (Chabrol and Muchnik 2011). This burden is not alleviated by finding products from a certain origin, as many GIs are heterogeneous, so further quality assessments and decisions are required. This is particularly the case with wine and spirits, but it is also important when choosing agri-food products. The aggregation that occurs through GI registration may mask, but does not strongly inhibit, a diverse range of expressions of quality by individual producers (Van der Ploeg 2008, 136). In this sense, GI is perhaps evolving to be an alternative form of requalification worldwide that is a counterweight to more mainstream processes of standardization and commoditization under globalization. However, without consumers exhibiting sufficient aptitude to differentiate products, and thereby justify higher value, even GIs will tend toward internal homogeneity or require additional layers of certification to perpetuate difference.

In the post-2000 era, governments adopting GI policies are also presented with an opportunity to leverage the logistical support (financial, technical and legal) and the trade benefits that the European Union has helped enshrine in TRIPS. As discussed in the context of marquee GI products, a new GI policy must at least have the basic instruments necessary to promote and register a few internationally oriented, EU-approved products but otherwise can be molded and implemented in ways that suit each domestic context. The advent of GI creates sudden new space for tailoring development to specialized producers, and much of the palpable excitement in Asia relates to this fact, but ultimately the flexibility to design their GI policy to "cater to domestic needs" can be interpreted by governments as an opportunity to calibrate their agricultural sector more to the neoliberal world order. In this process, if the ideals on which GI is based become suppressed, these new GI policies will be hollowed out and do little to provide an alternative to neoliberal market capitalism.

Notes

1 Organization for an International Geographical Indications Network (oriGIn), Strengthening International Research on Geographical Indications (SINER-GI), World Intellectual Property Organization (WIPO), Agence française de développement (AFD)

2 Place of origin certifications for alcoholic beverages are governed separately by the Japanese National Tax Agency
3 Personal communication, Nara, 21 August 2015.
4 Personal communication with representatives of the following products: Nishio matcha (Aichi), 22 January 2018; Hatcho miso (Aichi), 29 January 2018; Kobe beef (Hyogo), 7 March 2018.
5 Personal communication, Phnom Penh, 12 September 2017.
6 The financial burden of registering GIs is highly contextual. There are estimates of around US$1 million for the registration of the first two Cambodian GIs (Khmer Times 2015), but these cases were highly promoted for export. Notwithstanding, most of the Japanese GI registrations have been completed with minimal costs for registration (officially around US$800 for an indefinite GI registration) and administration due to the presence of preexisting producer groups (which play roles far beyond managing the GI).
7 Palm sugar is most commonly used as a paste but occasionally also in hardened blocks. Historically, it has not been granulated. Regarding pepper, Khmer cooking calls for either fresh whole corns or finely ground dust, which are distinct from the common product forms for Kampot pepper – coarse grinds or dried whole corns.
8 Personal communication: Phnom Penh, director of the DIP, Ministry of Commerce, 23 March 2018; Kratie province, head of the Koh Trung Eco-Tourism Community, 15 September 2017.
9 Personal communication, Okazaki-Aichi, 29 January 2018.
10 Personal communication, Nara, 21 August 2015; by telephone, 20 September 2015.
11 Opening speech by Chhim Vichara, head of the Battambang Provincial Department of Agriculture, Forestry and Fisheries. Presentation and Discussion on Research Results of Pursat-Battambang Orange Production and Market Linkage Project, 9 October 2017.
12 Personal communication with orange farmers in Pursat and Battambang provinces: 10–11 October 2017, 17 February 2018.
13 In 2017, the farmgate price of generic black pepper in Cambodia was US$3–4 per kg, whereas GI Kampot pepper farmers received US$15 per kg. Retail prices range from US$90 to US$300 per kg, depending on the quantity and form factor.
14 Personal communication, Kampot province, 5 September 2017
15 One notable exception here is the prominent Cambodian owner of Ngov Heng, who produces GI certified pepper as well as other uncertified products of origin, such as Kampot fish sauce.

References

Addor, Felix, Nikolaus Thumm, and Alexandra Grazioli. 2003. *Geographical Indications: Important Issues for Industrialized and Developing Countries*, 74. Seville: Institute for Prospective Technological Studies.

Allaire, Gilles. 2012. "The Multidimensional Definition of Quality." In *Geographical Indications and International Agricultural Trade: The Challenge for Asia*, edited by L. Augustin-Jean, H. Ilbert and N. Saavedra-Rivano, 71–90. London: Palgrave Macmillan.

Augustin-Jean, Louis. 2012a. "Introduction: The Globalization of Geographical Indications: The Challenge for Asia." In *Geographical Indications and International Agricultural Trade: The Challenge for Asia*, edited by L. Augustin-Jean, H. Ilbert and N. Saavedra-Rivano, 1–16. London: Palgrave Macmillan.

Augustin-Jean, Louis. 2012b. "Standardization vs Products of Origins: What Kinds of Agricultural Products Have the Potential to Become a Protected Geographical Indication?" In *Geographical Indications and International Agricultural Trade: The Challenge for Asia*, edited by L. Augustin-Jean, H. Ilbert and N. Saavedra-Rivano, 48–70. London: Palgrave Macmillan.

Belletti, Giovanni, Andrea Marescotti, and Jean-Marc Touzard. 2017. "Geographical Indications, Public Goods, and Sustainable Development: The Roles of Actors' Strategies and Public Policies." *World Development* 98: 45–57.

Blakeney, Michael. 2015. "Geographical Indications: An Introduction to the Literature." In *Geographical Indications*. Vol. 1, *Critical Concepts in Intellectual Property Law*, edited by M. Blakeney, xiii–lii. Cheltenham, UK: Edward Elgar Publishing.

Boisard, Pierre. 2003. *Camembert: A National Myth*. Berkeley and Los Angeles: University of California Press.

Bramley, Cerkia, Estelle Biénabe, and Johann Kirsten. 2009. "The Economics of Geographical Indications: Towards a Conceptual Framework for Geographical Indication Research in Developing Countries." In *The Economics of Intellectual Property: Suggestions for Further Research in Developing Countries and Countries with Economies in Transition*, 109–41. Geneva: World Intellectual Property Organization.

Calboli, Irene. 2017. "Geographical Indications Between Trade, Development, Culture, and Marketing: Framing a Fair(Er) System of Protection in the Global Economy?" In *Geographical Indications at the Crossroads of Trade, Development, and Culture: Focus on Asia-Pacific*, edited by I. Calboli and N. L. W. Loon, 3–22. Cambridge: Cambridge University Press.

———. 2006. *Expanding the Protection of Geographical Indications of Origin Under TRIPS: Old Debate or New Opportunity?* Milwaukee, WI: Marquette University.

Chabrol, Didier, and Jose Muchnik. 2011. "Consumer Skills Contribute to Maintaining and Diffusing Heritage Food Products." *Anthropology of Food* 8.

Cheng, Sokhorng. 2017. "Pepper Sector Shaken by Boom." *Phnom Penh Post*, August 17.

De Launey, Guy. 2005. "Cambodia Targets Organic Market." *BBC News*, October 17.

de Gaudemar, Matthieu. 2016. "Cambodian Producers Urged to Build Effective Brands." *Phnom Penh Post*, September 16.

Direction de l'Agriculture et du Commerce. 1899. *Les Plantations de Poiviers au Cambodge en 1899*, 18. Saigon: Gouvernement Général de L'Indo-chine.

Heath, Christopher. 2017. "How Would Geographical Indications from Asia Fare in Europe." In *Geographical Indications at the Crossroads of Trade, Development, and Culture: A Focus on Asia-Pacific*, edited by I. Calboli and N. L. W. Loon, 186–211. Cambridge: Cambridge University Press.

Hughes, Justin. 2017. "The Limited Promise of Geographical Indications for Farmers in Developing Countries." In *Geographical Indications at the Crossroads of Trade, Development, and Culture: A Focus on Asia-Pacific*, edited by I. Calboli and N. L. W. Loon, 61–86. Cambridge: Cambridge University Press.

IIP. 2011. *Working Paper on the Legal Survey of Geographical Indication Based on Preliminary Research Findings*. Tokyo: Institute of Intellectual Property (in Japanese).

Källander, Inger, and Gunnar Rundgren. 2008. *Building Sustainable Organic Sectors*. Bonn, Germany: International Federation of Organic Agriculture Movements (IFOAM).

Kerr, William A. 2006. "Enjoying a Good Port with a Clear Conscience: Geographic Indicators, Rent Seeking and Development." *The Estey Centre Journal of International Law and Trade Policy* 7 (1): 1–14.

Khan, Sophirom. 2016. "Kampot Pepper Successfully Registered as GI in EU." *Agence Kampuchea Presse*, February 23.

Khmer Times. 2015. "Agro Brand-Building Tool Has Benefits, and Drawbacks." March 19.

Kitagawa, Takako. 2005. "Kampot of the Belle Époque: From the Outlet of Cambodia to a Colonial Resort." *Southeast Asian Studies* 42 (4): 394–417.

Kojo, Daisuke. 2006. "The Importance of the Geographic Origin of Agricultural Products: A Comparison of Japanese and American Approaches." *Journal of Environmental and Sustainability Law* 14 (2): 275–321.

Lin, Lin, and Pei-Chuan Mao. 2015. "Food for Memories and Culture – A Content Analysis Study of Food Specialties and Souvenirs." *Journal of Hospitality and Tourism Management* 22: 19–29.

March, Elizabeth. 2007. "Geographical Indications: From Darjeeling to Doha." *WIPO Magazine*, 8–11.

Passeri, Stephane, and Channy Chheng. 2016. *A New Milestone for the Development of Geographical Indications (GI) in Cambodia. Press Release.* Phnom Penh, Cambodia: Food and Agriculture Organization of the United Nations (FAO).

Port, Kenneth L. 2014. "'Regionally Based Collective Trademark System' in Japan: Geographical Indicators by a Different Name or a Political Misdirection?" *Cybaris* 6 (2): 1–56.

Prak, Sereyvath. 2010. "Kampong Speu Palm Sugar, Cambodia." In *Quality Linked to Geographical Origin and Geographical Indications: Lessons Learned from Six Case Studies in Asia*, edited by A. Lecoent, E. Vandecandelaere and J. J. Cadilhon, 37–55. Bangkok: Food and Agriculture Organization of the United Nations (FAO).

Raustiala, Kal, and Stephen R. Munzer. 2007. "The Global Struggle Over Geographic Indications." *European Journal of International Law* 18 (2): 337–65.

Sekine, Kae, and Alessandro Bonanno. 2017. "Geographical Indication and Resistance in Global Agri-Food: The Case of Miso in Japan." In *Resistance to the Neoliberal Agri-Food Regime: A Critical Analysis*, edited by A. Bonanno and S. A. Wolf, 106–19. London: Routledge.

Société de géographie commerciale de Bordeaux. 1900. *Le Poivriers du Cambodge.* Bulletin: Bordeaux.

UNCTAD. 2015. *Myanmar's Unique Products to Benefit from "Geographical Indication" Promotion and Protection, UNCTAD Workshop in Yangon Told.* Geneva: United Nations Convention on Trade and Development.

UNESCAP. 2002. *Organic Agriculture and Rural Poverty Alleviation: Potential and Best Practices in Asia.* Bangkok: United Nations Economic and Social Commission for Asia and the Pacific.

Van der Ploeg, Jan Douwe. 2008. *The New Peasantries. Struggles for Autonomy and Sustainability in an Era of Empire and Globalization.* London: Earthscan.

Van der Ploeg, Jan Douwe, and Dirk Roep. 2003. "Multifunctionality and Rural Development: The Actual Situation in Europe." In *Multifunctional Agriculture: A New Paradigm for European Agriculture and Rural Development*, 37–53. Hampshire, UK: Ashgate.

van Ittersum, Koert, Math J. J. M. Candel, and Matthew T. G. Meulenberg. 2003. "The Influence of the Image of a Product's Region of Origin on Product Evaluation." *Journal of Business Research* 56 (3): 215–26.

3 The impact of geographical indications on the power relations between producers and agri-food corporations

A case of powdered green tea matcha in Japan

Kae Sekine

Introduction

In recent years, a number of Asian countries have introduced new legislation on geographical indications (GIs). The introduction of these measures coincided with the increased adoption of trade agreements and a more general tendency toward market liberalization (Augustin-Jean, Ilbert, and Saavedra-Rivano 2012; Calboli and Wee Loon 2017, Piatti and Dwiartama this volume; Feuer this volume). Following this trend, Japan introduced the Regionally Based Collective Trademark (hereinafter Collective Trademark) system in 2006 and the sui generis GI system in 2015. This sui generis system was based on the European Union schemes of GI (Augustin-Jean and Sekine 2012; Sekine and Bonanno 2018). While there has been an increasing number of studies on GI in recent years, GI systems in Asian countries remain relatively understudied compared to their European counterparts.

Although GI appears to be a desirable tool to protect local agriculture from the negative impacts of international market competition, revitalize rural communities and support traditional practices and food cultures, it also tends to be enthusiastically accepted as a tool to expand economic opportunities and increase exports.[1] The coexistence of these two contradicting dimensions of GI – protectionist and pro-market – can be identified as a basic feature of GI in the neoliberal globalization era (Bonanno this volume; Bowen 2015; Sekine and Bonanno 2018).

Over the last decade, international institutions, such as FAO and CFS (the Committee on World Food Security), have actively promoted new schemes to connect smallholder agricultures with "sustainable value chains" to achieve the Sustainable Development Goals (SDGs) (Kay 2016; Food and Agriculture Organization of the United Nations 2010, 2016, 2018; Neven 2014; Vandecandelaere 2010). In this context, GI is considered as one of the innovative tools that could alter conventional market systems and provide socially, economically and ecologically desirable alternatives. These favorable attitudes of international institutions about GI are supported by the scientific community (Barham and Sylvander 2011; Bowen 2015; Echols 2008; Van Caenegem and Cleary 2017).

From this perspective, the positive economic impacts of GI food systems are often seen as reaching all actors, including subordinated groups. In addition, the objective of territorial development is considered to be shared evenly by dissimilar actors, such as smallholders and agri-food corporations. However, in the rural realities, asymmetric power relations among GI actors coexist with competition, tensions and occasional conflicts (Nizam; Renard and Domínguez Arista; Weiler this volume). This situation raises the question about how the systematization of GI policies influence existing power relations among these actors. More specifically, will it lead to more equal or democratic relations, or will it instead advantage one party? If GI can contribute to mitigating economic power gaps among actors and co-constructing sustainable territorial development, what are the conditions that will enable this process to occur? Employing the case of GI Nishio matcha, a special sort of powdered green tea, produced in Nishio, Aichi Prefecture, in Japan, this chapter examines whether GI systems influence power relations and whether these systems contribute to the democratization of food systems. The relevance of the case study rests on the coexistence of local small-scale tea producers who employ GIs to withstand the consequences of agricultural crisis and transnational agri-food corporations that attempt to use GIs to drive matcha exports.

The chapter's first section provides the framework and methods for analyzing power relations in GI food systems, while the second presents a brief history of the development of GI in Japan. The third section illustrates the evolution of the matcha commodity chain and its power relations. Employing the case of GI Nishio matcha, in the fourth section, the power relations in this commodity system are analyzed. The chapter concludes with a discussion of the conditions that would allow GI to be an effective tool for the democratization of food systems and the roles of appropriate public policies.

Framework and methods

Uneven power relations in food systems are one of the most studied themes in the sociology of agriculture and food (Bonanno and Constance 2008; Sekine and Bonanno 2016; Wright and Middendorf 2007), food system studies (Niiyama 2001) and management studies (Torrès 2005). These studies focus on strategies and behaviors of powerful agri-food corporations and the resistance of subordinated groups against corporate power. Among these studies, sociological studies on contract farming contributed to disclosing the vertical power relations between producers and agri-food corporations (Burbach and Flynn 1980; Glover and Kusterer 1990; Little and Watts 1994; Magdoff, Foster, and Buttel 2000; Vellema 2002). They point out that degrees of vertical coordination in agri-food chains can vary according to market structures and producers' ability to resist integration and subordination.

From these studies, the chapter employs the following six dimensions to analyze vertical power relations between producers and agri-food corporations in GI food systems: (1) market structure – concentration, size of corporations and the existence of competitors and producers' organizations; (2) key elements that define qualities of GI products and the actors who control them; (3) the degree of producers' subordination, their autonomy in decision-making regarding production and their degree of financial control; (4) the evaluation of quality – measures for evaluation, actors involved in evaluation and a licensing system for evaluators; (5) forms of transactions – auction or contract (written or customary) terms of transaction; and (6) forms of pricing and the actors who influence prices in each transaction. To explore possible trajectories of sustainable territorial development in GI areas, the chapter also considers the following points: (7) common visions and strategies of territorial development among actors; (8) working conditions; (9) ecological sustainability; and (10) consumer awareness and the appreciation of GI products.

As most of the data are not available from official statistics, I employ a qualitative case study data analysis. The chapter is based on the semi-structured interviews conducted from 2017 to 2018 (see the list of interviews). In addition to the use of primary data, I also examined existing articles, documents and public and private statistics.

Development of geographical indication systems in Japan

After the Second World War, Japan experienced drastic social reforms followed by rapid economic growth and urbanization, resulting in the reshaping of rural lives, rural-to-urban out-migration and the shrinking of the agricultural sector (Sekine and Bonanno 2016). When Japan began to embrace neoliberal policies in the 1980s, market liberalization was imposed at the expense of domestic agricultural producers and rural communities, which consequently experienced a series of crises.[2] However, given that the number of agricultural producers has fallen by 30 percent since the mid- 2000s, more than 70 percent of farm owners are 60 years old or older and about 10 percent of farmland has been abandoned, we can consider Japanese agriculture to be in permanent crisis. In 2016 and from a food sovereignty perspective, Japanese agriculture satisfies only 38 percent of its domestic caloric demand. The crises of agriculture and rural communities are identified as some of the country's most urgent political challenges (MAFF 2015).

To alter the situation, GI appeared as a promising tool for the revitalization of Japanese agriculture that remained characterized by small operations, traditional culture and food heritages (Augustin-Jean and Sekine 2012; Sekine and Bonanno 2018). However, the adoption of GI in the country was not a linear process, reflecting international controversy over the legitimacy of GI under the WTO regime and competing ministerial strategies over GI. First, following the TRIPS Agreement, in 1995, the Japanese government revised the Liquor Tax Law and created a GI system for alcoholic beverages (National Tax Agency

2018). As of 2018, seven alcoholic beverages are registered as GIs under this law. Then, in 2005, JFIA (Japan Food Industry Association), under the direction and financial support of MAFF (Ministry of Agriculture, Forestry and Fisheries), established a voluntary certification system for GI named *Honbano Honmono* (authentic food of origin). As of 2019, forty-six agri-food products have been certified under this initiative (Honbano Honmono Brand Promotion Organization 2018, Interview with TTJ) (see Table 3.1). Second, and following the US position on GI, the JPO (Japan Patent Office), under the aegis of METI (Ministry of Economy, Trade and Industry), revised the trademark law to create the Regionally Based Collective Trademark in 2006. As of 2019, it has certified 646 agri-food and non-agri-food products (JPO 2019, Interview with JPO). In 2015 and after Japan and the European Union initiated negotiations on the Economic Partnership Agreement, the Japanese government created a European sui generis GI system to encourage exports to the EU market (Interview with MAFF). As of 2019, seventy-three agri-food products have been certified under this legislation (MAFF 2019).

To summarize the features of Japanese GI systems, I point out that (1) the coexistence of several GI systems reflects the political and economic position of the Japanese government floating between the United States and the European Union. (2) The government did not adopt the Protected Denomination of Origin (PDO) category under the sui generis GI system, leaving room for the certification of GI products made from imported raw materials. While it limits attempts to promote local agriculture and territorial development, it benefits countries that export raw materials and feeds to Japan. (3) The number of products certified under GI systems, especially those of the sui generis GI system, has increased rapidly. EU countries have registered 1,382 GIs and 1986 alcohol beverages since 1992, whereas Japan has totaled 646 Collective Trademarks since 2006 and seventy-three GIs since 2015. As Kimura and De Francesco (2017) mentioned, the Japanese GI system includes not only products such

Table 3.1 Public GI Systems and Private Territorial Labeling in Japan

Year	*GI for Alcohol*	*Collective Trademark*	*GI for Agri-food*	*Honbano Honmono*
	1995	*2006*	*2015*	*2005*
Ministry/Agency	National Tax Agency	Japan Patent Office (Ministry of Economy, Trade and Industry)	Ministry of Agriculture, Forestry and Fisheries	Japan Food Industry Association (JFIA)
No. of Products	10 (alcohols)	646 (11 alcohols, 306 non-food)	73 (6 non-food)	46 (only agri-food)
Legislation	liquor tax law	trademark law	sui generis gi law	–

Source: Author elaboration on data from National Tax Agency 2018, JPO 2019, MAFF 2019, Honbano Honmono Brand Promotion Organization 2018.

as Kobe beef, which is highly acclaimed internationally, and locally important, but also domestically unknown products such as Edosaki Pumpkin. To seek potential GI products, MAFF actively encourages agri-food producers to apply to its GI registration systems. (4) In contrast to the supportive attitude of MAFF toward GI applications, the ministry's ability to support actors' applications seems underdeveloped, mostly due to a lack of familiarity with, and understanding of, the European-origin GI system. When disagreements and conflicts occur among actors over GI registration, which happens quite often in many countries, MAFF tends to approve less stringent standards for GI products in order to include as many actors as possible (Japan Agricultural News 2016). This lenient attitude of MAFF, however, has caused serious disputes over GI registration such as in the case of "Hatcho Miso" (Sekine and Bonanno 2018).

Matcha commodity chain

Brief history of matcha

Today matcha is consumed in many major cities around the world, often in the form of matcha lattes in coffee shops, and has become a part of modern life, particularly among health-conscious consumers. What is matcha? Matcha means "powdered (green) tea" in Japanese and usually is considered a special sort of powdered green tea produced in Japan (Flint and Kavaliunas 2017). The raw material of matcha is green tea leaves harvested from *Camellia Sinensis*, which can also be transformed into black tea and oolong tea through different processes of fermentation. According to a voluntary standard defined by a Japanese tea business circle, matcha is produced from special green tea leaves called Tencha and milled using stone tea mortars. According to this definition, Tencha consists of tea leaves grown in the shade and processed without rolling. Therefore, under this definition, not all milled green tea can be considered matcha. However, this voluntary standard is not mandatory and is not always respected (Interview with Nishio Tea Cooperative).

As matcha contains antioxidants such as catechin and polyphenols, it is proved to be healthy and a protection against cardiovascular diseases, high cholesterol and blood pressure (Flint and Kavaliunas 2017). It is also considered a detoxicant of blood, which stabilizes blood sugar levels (Du et al. 2012; Flint and Kavaliunas 2017). It further contains 140 times more epigallocatechin gallate (EGCG), another antioxidant, than green tea; 25 percent less caffeine than coffee; and five times more L-theanine – a substance that decreases stress – than green tea (Flint and Kavaliunas 2017).

Historically, matcha was used for medicinal purposes. It became a popular drink in the eighth century in China. In the twelfth century, a Buddhist monk brought matcha making methods and tree seeds from China to Japan (Channell 2016, Flint and Kavaliunas 2017). Until the collapse of the Japanese feudal order in the mid-19th century, the matcha culture associated with the tea ceremony was an essential component of social life and diplomacy among the Japanese elite. In this way, the tea ceremony came to be a centerpiece of Japanese culture. However, from the mid-19th century, when Japan began to westernize, the tea ceremony gradually lost its symbolic status. After the Second World War, the rapid westernization of the Japanese lifestyle and diet led to a further fall in the consumption of matcha. Consequently, by 2016, ceremonial-grade matcha accounted for only 3.3 percent of green tea production in Japan (MAFF 2017a).

The transformation of the Japanese lifestyle also affected the tradition of drinking green tea, called Sencha, which is the most consumed green tea in Japan. Beginning in the late 20th century, Sencha served in traditional teapots was rapidly replaced by green tea sold in PET (polyethylene terephthalate or recyclable) bottles supplied by the beverage industry, often using imported green tea leaves (Ikegami 2011). To adapt to this transformation and command higher prices, green tea producers shifted their production from Sencha to Tencha, which created a boom in the raw material for matcha. From the late 20th century, most of the increased matcha supply has been absorbed by the food industry as culinary-grade matcha, particularly by the confectionary industry but also by the medicinal and cosmetic industries (Interview with Aiya). In the first decade of the 2000s, Starbucks began the sale of "matcha au laits" in Japan and, by the 2010s, its popularity expanded to New York, Paris and other major cities in the world. In the United States, individual coffee shops in San Francisco and New York also boosted the recognition of matcha in the country, especially among young generations (Interview with Matchabar). Social media, such as Instagram, and coverage of the matcha boom in the *New York Times* and *LA Times* supported this trend (Flint and Kavaliunas 2017).

The increasing demand for matcha-based products has led to a proliferation of substandard powdered green tea on the market that should not rightly be labeled *matcha*. This would suggest that there is a need to establish rigorous labeling standards to differentiate matcha. As there is no international legal definition of matcha, quasi-matcha threatens not only authentic matcha producers but also consumers and matcha culture. Because consumers cannot verify whether the matcha comes from Japan, whether it is made from Tencha or whether it does not contain artificial dye (Flint and Kavaliunas 2017), the incentives for fraud are high. To overcome this ambiguity, there are initiatives to regulate matcha labeling at the international level (e.g., International Organization for Standardization (ISO) on matcha) and national level (Japan Agricultural Standard, or JAS, in Japan and Chinese National Standard on matcha).[3]

Power relations among actors

Understanding the entirety of the matcha commodity chain is nearly impossible, because it retains many characteristics of merchant capital that dominated the pre-Meiji trading system that makes the collection of the business data highly problematic. Furthermore, the ill-defined boundary between green tea and matcha makes the collection of accurate official statistics on matcha difficult if not impossible.

First, while wholesale markets for green teas exist all over Japan, there are only a few wholesale markets where matcha auctions are carried out. In the matcha commodity chain, producers sell fresh tea leaves to primary processors, who often operate tea farms and produce dried tea leaves, called aracha. Then primary processors provide their products to secondary processors, who further refine, blend and mill dried tea leaves to make matcha. The transactions between producers and primary processors and between primary and secondary processors are based on contracts – written or unwritten, with long-standing relations often lasting over generations – and negotiation-based pricing. Some of the ancestors of matcha processors were matcha purveyors for tycoons and the emperor's family during the Tokugawa shogunate. To ensure the transaction of a limited quantity of a high-price product, producers and processors retained the customs of traditional transactions even after the democratization of the agri-food sector with the establishment of public wholesale markets and agricultural cooperatives in the 20th century.

Second, precise figures on matcha production are not available, due to the lack of an agreed-on definition of matcha and the existence of multiple institutions that oversee the sector. As mentioned, producers frequently harvest tea leaves for matcha, sencha and black tea from the same tea trees. Producers are able to decide which types of tea to produce every year, or even every harvest (Interview with Aichi Prefecture).[4] Moreover, as there are no standards for matcha, there is no institutional control on what producers and processors can call matcha. For this reason, official statistics are available for tea production as a whole, but they are lacking for matcha specifically. Furthermore, the statistics on matcha provided by the sector business association only covers production by its members.

Given that reliable statistics are unavailable to provide a reliable picture of the market structure of matcha, I conducted empirical case studies to analyze the features of this commodity chain. Okura (2004) states that the primary processing of tea leaves requires heavier investment than that of coffee, resulting in smallholder tea producers' dependency on processors. In addition, as with other types of tea, in the matcha commodity chain, the evaluation of tea leaves' quality is dominated by tea blenders, called tea masters, who work for primary and secondary processors (Interview with Aoi, Nishio Tea Cooperative and Nishio City Tea Association). Even though natural resources, soil, climate and water are the most important influencers of tea qualities (Interview with Aoi), the market structure and transaction customs allow processors to dominate the commodity chain. Usually these tea blenders have licenses approved by the National Tea Industry Youth Organization, the appropriate voluntary business association. As the pricing of fresh and dried tea leaves depends on quality evaluations, processors take initiatives in pricing. In addition, the higher concentration of capital in the secondary processing sector, relative to the primary processing sector, results in secondary processors wielding greater power over negotiations on transactions and pricing than primary processors have (Interview with Nishio Tea Cooperative and Nishio City Tea Association).

Although secondary processors of matcha run smaller firms with less capital than the large processing, trading and retailing corporations, the secondary tea processors use their market share in matcha to maintain a strong bargaining position when interacting with much larger transnational corporations. Primary producers in Nishio, Aichi Prefecture, have employed GI to change this structure of the matcha commodity chain.

A case of Nishio Matcha

Nishio Matcha in transition

Nishio matcha was registered as the Regionally Based Collective Trademark Nishio Matcha in 2009 and the GI Nishio Matcha in 2017. As of 2017, 138 producers[5] operate 210 ha of tea farms located in Nishio (205 ha) and Anjo (5 ha), Aichi Prefecture. At the municipality level in Japan, Nishio ranks first in the production of matcha (479 Mt/year in 2016). The municipality accounts for 88.5 percent of the prefecture's matcha production. In 2016, Aichi Prefecture was the second largest supplier of matcha (21.0 percent of total production in Japan) after Kyoto (41.7 percent) and ahead of Shizuoka (18.3 percent) (MAFF 2017a).

Tea cultivation in Nishio began in 1271, when a Buddhist monk sowed tea seeds in his temple, mainly for medical use (Sakaguchi 2009). From the 1870s, tea production in Nishio expanded among local farmers for commercial purposes in the Inariyama Plateau, located on the Yahagi delta. This area has fertile soil with good drainage, availability of abundant pure water, fog from Yahagi River that protects tea leaves from sunshine and frost and moderate temperatures, which provides the ideal environment for tea production. In 1884, the first tea leaf producers' organization was established, and it produced high-grade green tea, in Gyokuro. To receive higher farmgate prices, the organization moved production from Gyokuro to Tencha in the 1920s. Today more than 96 percent of tea produced in Nishio is shipped for matcha production (Nishio Tea Cooperative 2016). From the mid- 1930s, Tencha production further extended in the area and Nishio became one of the most important matcha production sites in the country. Along with the growth of tea production, primary and secondary matcha processors emerged in the area. Today there are thirty-six primary processors and seven secondary processors in Nishio, including Aiya, the largest matcha transnational corporation in the world. The availability of granite for tea mortars, made from granite produced in Okazaki, the adjacent municipality of Nishio, and accessibility to the large market of Nagoya, the third largest city in Japan, also favored the growth of Nishio Matcha. In the late 1950s, Nishio Matcha received a national award and thereafter became famous for its quality in the matcha industry (Interview with Aoi).

The business environment facing tea producers was transformed in the 1960s when Japan experienced rapid economic growth and began to liberalize its agri-food markets. The government encouraged farmers to diversify their production to include more profitable commodities (Sekine and Bonanno 2016). As more farmers began to produce tea, including matcha, the prices of matcha collapsed (Sakaguchi 2009). To acquire new markets, matcha processors in Nishio promoted the use of matcha for beverages (e.g., matcha lattes) and Western confections (e.g., ice cream, chocolate and cookies). The culinary-grade matcha for these purposes contains less flavor and savory or umami taste, and it is bitter than the premium ceremony-grade matcha (Interview with Nishio Tea Cooperatives, Nishio City Tea Association and Aoi). Processed foodstuffs containing matcha became increasingly popular in the 1990s. In the 2000s, major food industries such as Starbucks, Nestlé and Häagen-Dazs achieved global success with matcha products, and Nishio Matcha became one of the most important suppliers of matcha to these corporations.

However, the name Nishio Matcha remained unknown among consumers, as it did not appear on packages or menus (Interview with Nishio Tea Cooperatives, Nishio City Tea Association and Aoi). Among consumers, Uji Matcha still remains the best-known and appreciated name for quality matcha because of its historical importance as the matcha supplier for the ancient capital of Kyoto, which is the heart of the tea ceremony culture. Therefore, until the 2000s, about half of Tencha produced in Nishio was shipped to Uji in Kyoto Prefecture to be mixed and sold under the Uji Matcha brand (Interview with Nishio City Tea Association and Aichi Prefecture). Like producers in Nishio, producers in Shizuoka, Nara and Shiga Prefectures also shipped their aracha to Kyoto to benefit from the Uji Matcha brand that is associated with matcha in general. However, after the 2004 strengthening of the green tea food labeling regulation that includes matcha, matcha that contains more than 50 percent of tea grown outside of Kyoto could no longer be labeled as Uji Matcha. Since this time, matcha producers have looked for tools to differentiate their matcha from others and survive in the highly competitive matcha market.

Nishio Matcha under two geographical indication systems

To respond to the two challenges facing Nishio Matcha, the growth of new competing production sites and enhanced labeling regulation, Nishio Matcha producers enthusiastically sought novel tools to improve the recognition of Nishio Matcha among consumers and boost its prices (Interview with Nishio Tea Cooperatives, Nishio City Tea Association and Aoi). In 2007, an inter-professional organization, Nishio Tea Cooperative, comprising associations of producers, processors and retailers, was established to obtain and manage the Regionally Based Collective Trademark Nishio Matcha. In 2009, Nishio Matcha was defined as "Matcha made from tea grown and processed by tea-mortars in Nishio City, Anjo City and Kira Town"[6] under the Regionally Based Collective Trademark. Since 2009, the "Nishio Matcha" Regionally Based Collective Trademark and its accompanying logo have appeared on the packaging of more than three hundred items, tremendously improving the recognition of Nishio Matcha among consumers. Today around 80–90 percent of

Nishio Matcha is processed in the Nishio area (Interview with Nishio City Tea Association). However, some ambiguity remained over the definition of Nishio Matcha, because it did not exclude the use of tea leaves grown without shades as raw material or the use of tea mortars made with non-local granite stones. As a result, throughout the 2010s and despite the increasing popularity of Nishio Matcha, the farmgate prices continued to decrease (Interview with Nishio City Tea Association and Nishio Tea Cooperative).

In 2015, when MAFF introduced the sui generis GI system and encouraged the Nishio Tea Cooperative to register Nishio Matcha, the cooperative, and particularly its producer members, embraced the opportunity to improve the quality and brand recognition of their product by eliminating the ambiguities associated with the definition of matcha. GI Nishio Matcha, registered in 2017, is defined as "Matcha made from tea leaves grown more than 25 days for first flush, more than 12 days[7] for second flush under shelf-type shades,[8] dried with Mikawa-style Tencha Drying Machines,[9] milled by tea mortars made by granite stones from Okazaki City,[10] in Nishio City and Anjo City" (MAFF 2017b).

Although the more stringent criteria seem favorable after decades of struggles to distinguish their product, actor expectations of GI vary. On the one hand, as these vigorous methods contribute to a more vivid color, less bitterness and a mellower and stronger umami taste of matcha, producers consider that the recent stabilization of farm gate prices is due to GI registration. They further expect that GI generate higher farmgate prices in the future (Interview with Nishio Tea Association). On the other hand, the transnational processors in Nishio operate their matcha factories in Kyoto and procure tea leaves from several regions of Japan to balance qualities and prices. The much stricter definition of GI Nishio Matcha puzzled corporations, because they are not necessarily interested in sourcing GI Nishio Matcha (Interview with Aiya, Aoi and Nishio Tea Cooperative). However, producers and processors share a common interest in protecting the GI from fraud and/or misuse. For instance, in 2017, several foreign corporations spuriously applied the Nishio Matcha trademark in China, Europe and the United States (Interview with Nishio Tea Cooperative).

Power relations in Nishio Matcha production

According to the framework that I have proposed, I analyze the power relations among actors in the Nishio Matcha commodity chain along six axes. (1) The market structure is highly concentrated at the secondary processing stage. Although there are 138 producers, including thirty-six producers/primary processors, there are only seven secondary processors, two of which are transnational corporations: Aiya and Aoi (Interview with Nishio Tea Cooperative, Aiya and Aoi). Aiya was established in 1888, employs 192 people and has US$273,000 in capital stock and counts on a global market share estimated at 30 to 40 percent in 2018. Aiya purchases 20 percent of total tea production for matcha in Nishio from 40 percent of producers in the area, and it mills the tea leaves with a thousand tea mortars (Aiya 2017; Interview with Nishio

Tea Cooperative). It operates seven subsidiaries, in total, in the United States, Germany, Austria, China and Thailand. Aoi was established in 1966, employs sixty people and holds US$182,000 in capital stock in 2018 (Aoi 2018). Aoi purchases tea leaves from about 20 percent of Nishio producers, who provide 30 to 40 percent of the raw material of Aoi's matcha production (Interview with Aoi). It operates a subsidiary in the United States. Producers sell 80 to 90 percent of their production to corporations, which hold local monopolies in the processing of Nishio Matcha. These matcha processing corporations are able to procure tea leaves from other regions of Japan and sell their products globally through diverse marketing channels. The producers, hoping to valorize local production, have a strong interest in the promotion and recognition of the Regionally Based Collective Trademark and GI Nishio Matcha. Accordingly, these transnational corporations are less engaged in regional development and in the welfare of local farmers. These relatively small but dominant two transnational matcha corporations supply matcha to major agri-food corporations such as Starbucks, Nestlé, Häagen-Dazs and Wholefoods Market (Interview with Nishio Tea Cooperative, Aoi and Matchabar). In 2017, about 20 percent, or 100 t, of Nishio Matcha was exported abroad.

(2) According to the processor Aoi, the key elements that influence the qualities of Nishio Matcha are (a) the soil, (b) climate, (c) timing and techniques of harvesting and d) varieties of tea trees. These factors all depend on the natural and human resources in the farming phase. (3) Since processors do not affect conditions in the farming phase, producers remain relatively autonomous compared to black tea plantations in developing countries. Producers' good access to financial resources, such as public subsidies and private loans, also contribute to their relative independence (Interview with Akahori). However, (4) because licensed tea masters working for processors dominate the evaluation of tea leaf quality and (5) because they follow written or customary contracts with producers stipulating that producers must sell all their production exclusively to processors, (6) processors take primary leadership in pricing. The form of pricing is reference pricing. This is the price that the leading corporation, Aiya, establishes through negotiations with its contractors and that serves as the benchmark price (Interview with Nishio Tea Cooperative). Nishio Tea Cooperative compared Coca-Cola, the price leader of coffee beans, with Aiya, the price leader of matcha globally. Producers state that even though they disagree with the prices proposed by processors, they cannot sell tea leaves to processors in other regions (Interview with Nishio City Tea Association).

From the case study, we can observe that GI systems do not necessarily change these market structures and transaction customs. However, there is room to exert influence on existing power relations among actors if these systems promote consumer recognition and willingness to choose GI products. A processor admits that organic-certified producers retain stronger negotiation power than processors because of the scarcity of organic tea leaves amid growing demand.

To envision sustainable territorial development, can GI systems contribute to the resolution of the social and ecological dimensions? (7) The relevance and necessity of promoting GI systems are shared among actors (Interview with Nishio Tea Cooperative, Nishio City Tea Association and Aoi). GI also contributed to an increase in participative attitudes and feelings of attachment among local residents to promote Nishio Matcha. For instance, there are several events associated with Nishio Matcha, such as food education in local school lunch programs and tea-harvesting training, the world's biggest tea party recorded in the Guinness World Records and agritourism that allows local guides to learn about Nishio Matcha.

(8) The working conditions of Nishio Matcha producers, especially for seasonal harvest workers, could be considered the most contentious issue. Even though 80 percent of the harvesting process is mechanized, the high ceremonial-grade matcha cannot be produced without handpicking. While fresh tea leaves fetch producers an average farmgate price of US$2.5 USD per kilogram, harvesters receive only US$1.2/kg (Interview with Nishio Tea Cooperative and Aoi).[11] As regular farm workers can harvest about 20 kg/day, their daily wage is estimated to be only US$24.[12] Compared to the average daily wages in manufacturing or the service sector (about US$73), it remains extremely low. This occupation is predominantly filled by local elderly female workers who are willing to work at these wage levels. Immigration of non-skilled workers is not permitted. The sustainability of Nishio Matcha depends on the willingness and availability of these harvesters, who share feelings of conviviality and attachment to harvesting. Additionally, several times a year, local students participate in the harvest on a voluntary basis. Accordingly, GI offers limited solutions to the improvement of working conditions of producers and, in particular, wage workers.

(9) From an ecological perspective, a direct positive impact could not be observed in this case study. As ecological requirements are not a mandatory component of GI systems, GI does not necessarily contribute to biodiversity or to the use of fewer agro-chemicals. Rather, as is usual in tea production, Nishio Matcha production is based on monoculture and the application of agro-chemicals. However, along with the trend to reduce agro-chemical inputs in recent years, producers in Nishio have also reduced agro-chemicals and increased organic manure (Interview with Nishio City Tea Association). Yet, as of 2018, there are no certified-organic producers carrying out Nishio Matcha production. To make sure that GI contributes to ecological sustainability, ecological requirements would need to be explicitly institutionalized in the GI system itself or be included in other environmental compliances.

(10) Can GI increase consumer awareness and an appreciation of GI products and their qualities? GI systems and their official logos can improve consumers' recognition of products and can be effective tools to inform consumers about their particular terroir, traditions, nutrition values, associated cultures and difficulties that actors face today. If GI could play a role of linking consumers to

GI territories and territorial actors in such a way, it would get one step closer to sustainable territorial development.

Conclusions

The case of GI systems in Japan indicates that trademark-based and/or sui generis legislation–based GI systems are adopted and can coexist in Asian countries. It also shows that they can be employed to overcome economic and social challenges, albeit in the context of limitations imposed by existing political and diplomatic conditions. The sui generis GI system shows higher potential to guarantee the quality of GI products and contribute to the creation of better product recognition. It also can add to improving the economic status of stakeholders in ways that are better than those offered by trademark-based GI systems. However, as indicated in the case of Nishio Matcha, even sui generis GI systems do not systematically and positively contribute to the democratization of the existing power relations. This is particularly the case in processed food industries that are dominated by export-oriented oligopolistic firms.

From a sustainable territorial development perspective, GI systems cannot automatically guarantee positive results. Therefore, conditions that would democratize food systems and contribute to the emancipation of subordinated groups require that alternative public policies be developed. These policies should incorporate not only measures in support of GI systems that could guarantee a more equitable redistribution of rent among stakeholders and the safeguarding of biodiversity and ecological sustainability but also institutional effectiveness in terms of significant institutional support for those involved.

Notes

1 Sui generis GI legislation suddenly became an urgent project when Japan began to negotiate the EU-Japan Economic Partnership Agreement in the 2010s. This was in the line with the Japanese government's policy to promote exporting agri-food products.
2 Japan is often considered as one of the countries that significantly subsidizes its agricultural sector. However, as indicated by Suzuki and Kinoshita (2017), this argument is misleading in that it is based on inaccurate data.
3 In May 2018, the Chinese National Standard on Matcha went into effect as the first world legal regulation of matcha (Tea Media Corporation 2018). While it refers to shade, it does not identify Japan as the origin of matcha.
4 In Japan, producers harvest tea leaves between one and five times a year.
5 Between the 1980s and 2010s, the total quantity of land cultivated for tea leaves remained constant. However, the number of tea producers sharply declined, to 138 from the original 500 in the same time period.
6 Kira was aggregated into Nishio in 2011.
7 Longer covering periods result in richer L-theanine and better taste. GI Nishio Matcha is the first matcha that defined the exact dates required for matcha.
8 To avoid direct contact between tea leaves and sunshades, shelf-type sunshades are preferred to direct-covering sunshades. It generates finer tea leaves with superior taste.
9 It is a locally developed special machine to dry fresh tea leaves with infrared rays and a five-layered drying stage.

10 Granite contains less heat as it is employed to mill tea leaves into matcha. The tea mortar revolves at a rate of sixty times per minute and mills only 40 g/hour to keep the flavor, taste and color of matcha. Other types of modern mills can produce 10 to 12.5 kg/hour.
11 Tencha is sold at US$28/kg on average (MAFF 2017a), and the matcha is sold at US$18/kg for culinary grade and US$360–540/kg for ceremonial grade (Interview with Aoi).
12 Though this is below the local minimum wage, piecework is not covered by minimum wage legislation.

Bibliography

Aiya. 2017. *Aiya: An Exclusive Manufacturer of Matcha Since 1888*. Nishio: Aiya.

Aoi. 2018. *Company Profile and History*. Accessed April 30, 2018. www.aoiseicha.co.jp/aoi/index.html (in Japanese).

Augustin-Jean, Louis, Hélène Ilbert, and Neantro Saavedra-Rivano, eds. 2012. *Geographical Indications and International Agricultural Trade: The Challenges for Asia*. New York: Palgrave Macmillan.

Augustin-Jean, Louis, and Kae Sekine. 2012. "From Products of Origin to Geographical Indications in Japan: Perspectives on the Construction of Quality for the Emblematic Productions of Kobe and Matsusaka Beef." In *Geographical Indications and International Agricultural Trade: The Challenges for Asia*, edited by Augustin-Jean, Louis, Hélène Ilbert and Neantro Saavedra-Rivano, 139–63. New York: Palgrave Macmillan.

Barham, Elizabeth, and Bertil Sylvander, eds. 2011. *Labels of Origin for Food: Local Development, Global Recognition*. Wallingford: CAB International.

Bonanno, Alessandro, and Douglas H. Constance. 2008. *Stories of Globalization: Transnational Corporations, Resistance, and the State*. University Park, PA: Pennsylvania State University Press.

Bowen, Sarah. 2015. *Divided Spirits: Tequila, Mezcal, and the Politics of Production*. Oakland, CA: University of California Press.

Burbach, Roger, and Patricia Flynn. 1980. *Agribusiness in the Americas*. New York: Monthly Review Press.

Calboli, Irene, and Ng-Loy Wee Loon, eds. 2017. *Geographical Indications at the Crossroads of Trade, Development, and Culture: Focus on Asia-Pacific*. Cambridge: Cambridge University Press.

Channell, Randi Soei. 2016. *The Book of Chanoyu: Tea … The Master Key to Japanese Culture*. Kkyoto: Tankosha Publishing.

Du, Guang-Jian, Zhiyu Zhang, Xiao-Dong Wen, Chunhao Yu, Tyler Calway, Chun-Su Yuan, and Chong-Zhi Wang. 2012. "Epigallocatechin Gallate (EGCG) Is the Most Effective Cancer Chemopreventive Polyphenol in Green Tea." *Nutrient* 4 (11): 1679–91.

Echols, Marsha A. 2008. *Geographical Indications for Food Products: International Legal and Regulatory Perspectives*. Amsterdam, NL: Kluwer.

FAO. 2018. *Strengthening Sustainable Food Systems Through Geographical Indications: An Analysis of Economic Impacts*. Rome: FAO.

———. 2016. *Innovative Markets for Sustainable Agriculture: How Innovations in Market Institutions Encourage Sustainable Agriculture in Developing Countries*. Rome: FAO.

———. 2010. *Linking People, Place and Products: A Guide for Promoting Quality Linked to Geographical Origin and Sustainable Geographical Indications*. Rome: FAO.

Flint, Jessica, and Anna Kavaliunas. 2017. *Matcha: A Lifestyle Guide*. New York: Dovetail Press.

Glover, David, and Ken Kusterer. 1990. *Small Farmers, Big Business: Contract Farming and Rural Development*. London: Macmillan Press.

Honbano Honmono Brand Promotion Organization. 2018. *Website: Certified Products.* Accessed April 16, 2018. https://honbamon.com/product/index.html (in Japanese).

Ikegami, Koichi. 2011. "The World Observed Through Pet Bottled Tea." In *Today's Agriculture and Food,* edited by Koichi Ikegami and Kosuke Harayama. Kyoto: Nakanishiya Publishing (in Japanese).

Japan Agricultural News. 2016. "A Year Later GI: Boost Prices, Challenges in Recognition toward Expansion of Registration." *Japan Agricultural News.* June 29, 2016 (in Japanese).

JPO. 2019. *The Guidebook of Collective Trademark.* Tokyo: JPO.

Kay, Sylvia. 2016. *Connecting Smallholders to Markets: An Analytical Guide.* Rome: International Food Security and Nutrition Civil Society Mechanism.

Kimura, Junko, and Edi De Francesco. 2017. "Value-adding on Agri-food Products by Geographical Indications: A Case of Japanese GI." *Working Paper of Japan Marketing Academy* 3 (12): 1–20 (in Japanese).

Little, Peter D., and Michael J. Watts. 1994. *Living Under Contract: Contract Farming and Agrarian Transformation in Sub-Saharan Africa.* Madison: The University of Wisconsin Press.

MAFF. 2019. *Website on Certified GI products.* Accessed February 12, 2019. www.maff.go.jp/j/shokusan/gi_act/register/index.html (in Japanese).

———. 2017a. *Circumstances of Tea.* Tokyo: MAFF.

———. 2017b. *Geographical Indication Registry of Nishio Matcha.* Tokyo: MAFF.

———. 2015. *Summary of the Basic Plan for Food, Agriculture and Rural Areas: Food, Agriculture and Rural Areas Over the Next 10 Years.* Tokyo: MAFF.

Magdoff, Fred, John Bellamy Foster, and Frederick H. Buttel, eds. 2000. *Hungry for Profit: The Agribusiness Threat to Farmers, Food, and the Environment.* New York: Monthly Review Press.

National Tax Agency. 2018. "Website on the List of GI Alcohols." Accessed September 30, 2018. www.nta.go.jp/taxes/sake/hyoji/chiri/ichiran.htm (in Japanese).

Neven, David. 2014. *Developing Sustainable Food Value Chains: Guideline Principles.* Rome: FAO.

Niiyama, Yoko. 2001. *Food System of Beef: A Comparative Analysis of Europe and the United States.* Tokyo: Nihon Keizai Hyoron Sha (in Japanese).

Nishio Tea Cooperative. 2016. *Nishio Matcha in Aichi.* Nishio: Nishio Matcha Cooperative.

Okura, Miwa. 2004. "Coffee, Tea, and Agribusiness." In *Food and Agribusiness Under Globalization,* edited by Shigeru Otsuka and Toyohiko Matsubara. Tokyo: Yuhikaku.

Sakaguchi, Kayoko. 2009. "Nishio Matcha." *CREC* 167: 57–68 (in Japanese).

Sekine, Kae, and Alessandro Bonanno. 2018. "Geographical Indication and Resistance in Global Agri-Food: The Case of Miso in Japan." In *Resistance to the Neoliberal Agri-Food Regime: A Critical Analysis,* edited by Alessandro Bonanno and Steven A. Wolf, 106–19. New York: Routledge.

———. 2016. *The Contradictions of Neoliberal Agri-food: Corporations, Resistance, and Disasters in Japan.* Morgantown, WV: West Virginia University Press.

Suzuki, and Junko Kinoshita. 2017. "The Lowest Self-Sufficiency Ratio Among Major Developed Countries." In *New Edition: Understanding Contemporary Agriculture, Food and Environment from Keywords.* Kyoto: Showado.

Tea Media Corporation. 2018. *Summary of Chinese National Standard on Matcha.* Accessed April 28, 2018. www.team edia.co.jp/blog_32/.

Torrès, Olivier. 2005. *L'affaire Mondavi: Mondialisation et terroirs.* Paris: Dunod.

Van Caenegem, William, and Jen Cleary, eds. 2017. *The Importance of Place: Geographical Indications as a Tool for Local and Regional Development.* Cham, SW: Springer.

Vandecandelaere, Emilie. 2010. "Geographical Origin and Indication Labels: Associating Food Quality with Location." In *Innovations in Food Labelling,* edited by Albert Janice. Rome: FAO and New Delhi: Woodhead Publishing.

Vellema, Sietze. 2002. *Making Contract Farming Work? Society and Technology in Philippine Translational Agribusiness.* Maastricht: Shaker Publishing.
Wright, Wynne, and Gerad Middendorf. 2007. *The Fight Over Food: Producers, Consumers, and Activists Challenge the Global Food System.* University Park: PA: Pennsylvania State University Press.

List of Interviews

Aichi Prefecture: February 2018
Aiya Co., Ltd.: July 2017
Akahori Ltd.: March 2018
Aoi Seicha Co., Ltd.: March 2018
Japan Patent Office: December 2017
Ministry of Agriculture, Forestry and Fisheries of Japan: December 2017
Nishio City Tea Association: January 2018
Nishio Tea Cooperative: January 2018 and March 2018
Terroir and Tradition Japan Co., Ltd.: December 2017

4 Provenance for whom? A comparative analysis of geographical indications in the European Union and Indonesia

Cinzia Piatti and Angga Dwiartama

Introduction

Although the history of geographical indications (GIs) dates back to before World War II, industrialization in Europe before and after World War II had initiated a period of major expansion of GIs, including outside of its birthplace. The growth of GIs was mainly the result of a targeted regional product recognition which developed in the shadow of a period of economic development characterized by cheap and industrially produced food to respond to the imperatives of the age of economic boom. GIs have since grown and developed to the point of becoming a quite attractive tool for new countries in recent times. The narrative of GI so far has been largely positive, with GI apparently offering a tool for the emancipation from ruthless capitalism by food producers – especially those small, marginalized or simply lacking the negotiating power of agri-food system actors in different parts of the world (e.g., Bowen 2010). But despite the benefits that usually accompany GIs, there are good reasons to question this narrative, especially for producers in developing countries. Increasingly, the limitations of GIs have become visible as new country data and new studies continue to explore the contours of power, geographic and market relations (Hughes 2017; Augustin-Jean 2012). Indeed, if GIs are positioned in the sociopolitical frame of neoliberalism, as this book so positions them, GIs offer a questionable antidote to extreme market-rule and economic liberalization, in which even vanguard agri-food marketing can be stymied by corporate domination.

If agri-food neoliberalism is epitomized by the corporate regime (McMichael 2005), with its associated logics responding to power and accumulation, then what is the realistic potential of GIs? GIs, which rely on trade and commerce even as they try to tame it (Bonanno, this volume), might not deliver the promise of recognition and reward or, even more importantly, change and disruption of established forms of trade power, market relations and hierarchical organization that are hegemonic under neoliberalism. And yet through institutional innovation, might there be any space left for disrupting this regime? To work through this, we begin this chapter by questioning whom this establishment of provenance benefits – that is, to which places and actors does the recognition of the origin of a food product deliver value and worth? The space where such institutional innovation might arise is perhaps in those liminal spaces where provenance has already garnered significant value but has been appropriated due to the (prior) absence of a GI system. To determine whether GI can play an emancipatory role against neoliberal capitalism, we propose a comparison of the European Union (EU) and Indonesia, because of the difference in timing of implementation of GI and the preexisting worldwide value of well-known place-of-origin products such as coffee between these two places.

We proceed through a three-dimensional analysis that comprises (1) sociocultural dynamics that highlight the embeddedness of GIs in the social fabric (Morris and Kirwan 2011; Rosin, Stock, and Campbell 2013), since the role of actors in creating – or maintaining – a GI goes well beyond the legal inception; (2) materiality embedded in the local ecology (Henry and Roche 2013), given the nature of food items as products of a specific place and ecology; and (3) power relations, which imply a power differential between actors (Carolan 2013a). These three dimensions allow us to draw attention to the different moments of the GIs process: before implementation, during market integration, through the moment of institutionalization and finally after the acquisition of GIs, showing the contingent process during which power differentials combine and crystallize. The redefinition of the value through the GI process makes us question how and for whom provenance has been designed and advocate for a more comprehensive understanding of its potential mechanisms.

Geographical indications

GIs are often perceived as a linearly positive mechanism, a win-win situation for producers and consumers, at both the individual and country levels, because they amplify market choice, differentiation and recognition. GIs are used to vindicate the virtuosity of what is "local," "traditional" or "authentic"; to gain extra value in a saturated market; to support rural development; and to furnish producers with legal protection. Some claim that GIs also have the power to become an alternative to the neoliberal regime (Quiñones-Ruiz, Xiomara, Penker, Vogl, and Samper-Gartner 2015). Increasingly, though, the way GI is implemented worldwide does not readily demonstrate this capacity. Indeed, even at the macro level, GIs may merely be extracting the residual value of exhausted markets, whose global system of production needs to be rescaled anyway to adapt to ecological and social pressures (Navé Wald and Douglas Hill 2016). Furthermore, the same vocabulary of corporate regimes is often used to describe GI. In contrast, some voices in agri-food circles argue that we are already too deeply immersed in a regime dominated by corporations (McMichael 2005), such that hegemonic production and distribution rules and practices can readily subsume most forms of challenge to the established regime. GI, as a facilitator of consumption, seemingly cannot escape a neoliberal regime constituted by corporatized trade, commoditization and the attendant political (de)regulation and social restructuring under the banners of liberalization and competition.

Some authors view GIs as spaces for engagement, whereby factors such as ecology and the heretofore limited powers of marginal actors might drive "impetus for reform, that is, for a new regime based on implicit rules of power, accumulation, and class" (Friedmann 2016) that will potentially steer the regime toward resilience and sustainability. If that is the case, then GI fits the debate, because its history belongs to a model of reconstituted marketization in the rush to conquer new profitable spaces – literally and metaphorically. This outcome is unclear, however, because of the accumulating failures of GI systems in the inclusivity of producers and governance in established GIs systems, even in the stronghold of the European Union but also increasingly in emerging systems such as Japan (Kizos et al. 2017) and Mexico (Renard and Domínguez Arista, this volume).

In newcomer GI countries, there are also failures in establishing legal frameworks for the implementation of the GI that protect new entrants, such as for coffee in Indonesia (Neilson, Wright, and Aklimawati 2018). This book features many other challenges, such as the cultural suitability of GI (Shtaltovna and Feuer, this volume), the efficacy of GI in upending power relations (Sekine, this volume) and the strategic management of heritage production (Niederle, Wilkinson, Mascarenhas, this volume). Perhaps most worrying – hence the topic taken up in this chapter – is whether GI may not only reinforce neoliberal regime strictures but also risk perpetuating chauvinistic relationships, such as colonial trade relations in the case of Indonesia.

Three dimensions of concern

As already introduced briefly, we maintain that there are three dimensions that pertain to GIs: first and foremost are the sociocultural dynamics inherent in a GI designation, which represent the fabric out of which a product emerges. These sociocultural dynamics highlight the embeddedness of food in a specific place where people tied to a specific culture live. Michael Carolan (2015) writes of food as something that is felt, practiced and performed in particular places and thus expresses belonging to people and places. Morris and Kirwan (2011) offer a good discussion of the concept of embeddedness, based on Mark Granovetter's (1985) conceptualization, highlighting the social and ecological relations intrinsic to such a concept, but they remind us to avoid simplifying the global/local binary. There are some specific cultures, values, practices that will make a product what it is, and the mechanism behind GIs is a way to codify and standardize them. People involved can be peasants, farmers and small artisans but also large groups organized around managerial tenets and conspicuous assets. Each of course contributes to the contemporary understanding and development of GIs but each has a different weight in the final result of obtaining a GI and the economic rewards that may arise. This means recognizing that

GIs offer a valid global model for reclaiming worth only when organized to account for specific places and people, because the role of actors in creating – or maintaining – a GI goes well beyond the legal inception. Based on our cases, we do not see this is happening as popularly expected.

This connects directly to the second dimension, which is the materiality embedded in the local ecology, because the assembling of different elements means involving a range of elements necessary to make a product worth the GI protection (Murray Li 2007). Regarding embeddedness, GIs should be the expression of specific ecologies, and as such, they would be made of different elements and actors that respond to different (eco)systems. Importantly, these elements should also be material or tangible, in our case foodstuff, which are nonetheless able to "exert their own agency through their ability to jump into trajectories that are neither foreseeable nor easily controllable" (Henry and Roche 2013, 199). In other words, because these materials change the very understanding of a place as much as they are also influenced by that place, they create an endogenous relationship between the environment and the product. Of course, whether the taste of a product reputedly associated with a terroir is universal and absolute is highly debatable. Here again, the recognition accorded to a product and its taste is endogenous: it is both a matter of "education" (as distant consumers must do) or "getting used" to a taste (Carolan 2013a), as a matter of imposing expectations and norms of quality upon others. This may then raise the question of who should impose a specific "universal" flavor and who would be qualified to judge a product (and decree the success or not of); thus, it runs counter to a logic of standardization that belongs to neoliberalism (see Busch 2011; Loconto and Busch 2010).

The third dimension is the power differential between actors at both intra- and inter-levels of analysis – that is, within a specific country for interested actors or between different nations – which may lead to the reconstruction of geopolitical paradigms. At the intra-level, we refer to the different individuals or groups domestically that can be involved in establishing or maintaining a GI and might have different understandings of a product and its features. At the inter-level, we refer to different countries negotiating the recognition of different GIs in their respective territories and trying to harmonize or simplify recognition while also competing for a similar high-end market. Here are the numerous international trade negotiations that clearly demonstrate the asymmetrical power relations (skewed in favor of Europe) for gaining protection of different products while also working to streamline mutual recognition. Carolan (2013b, 146) writes of difference power as a counterforce to dominant power, offering the perspective of a "multitude of performances from below," which in our perspective can work to alter what have become stabilized and often unescapable hierarchies. In the case of GIs, the power differential dimension highlights the relationships developing between actors and, at a higher level of commercial negotiations, involves a revision of trade relationships and geopolitical power.

Methods

While the narrative provided for the GI case in the European Union mainly originated from one of the authors' literature review and secondary works, some of the data for Indonesia's case study were taken from interviews with stakeholders in the GI community groups (*Masyarakat Pelindung Indikasi Geografis*, or MPIG) carried out between July and October 2017, in addition to desk evaluations of existing government reports and GI documents. Fifteen informants, which included local government officials, farmers and traders, were interviewed to understand their views on the importance of GI for three products, namely *sumedang mole* tobacco, *cilembu* sweet potato and *java preanger* coffee in Sumedang Regency of West Java Province (unpublished data).

Results and discussion

Geographical indication in the European Union and Indonesia: a comparison

The European Union and Indonesia represent two different forms of GI implementation and functioning. The difference between established and emerging GI systems is obviously rooted in preexisting conditions, namely for Indonesia in its colonial past and for the European Union in setting much of the initial regulations and market framework for GI. This heavily impacts the ability of Indonesia and its producers to implement GI in a way that disrupts the neoliberal market arrangements to its benefit. We will explore the central differences between these two regions and highlight how GI, as currently structured, is unlikely to play an emancipatory role in Indonesia.

First of all, the European Union is a conglomerate of states, which established the original model of GI and can rely on experience and a shared perspective of single products and their protection. Indonesia is instead a single nation-state, albeit the world's fourth most populous country, which has recently seen an upsurge in interest in GI but which lacks the same historical precedents for establishing a GI system. Second, the European Union draws on a rich variety of products, including wine, cured meats, vegetables, fruits and processed food, which correspond to the different kinds of legal categories, means and actions for obtaining protection. Indonesia has so far principally concentrated its efforts on a single product, namely coffee, which dominates the list of products that have been awarded protection. This imbalance, which coincidentally depends on structural conditions persisting from the colonial past of Indonesia, results in a systematic bias toward commodities that fit a wide, international demand. Third, the European Union relies on a multi-niche production and a structured bureaucratic and legal apparatus that ensures GI procedures and raises the chances of success, whereas Indonesia has so far relied on producing commodities and increasing domestic demand more generally but is still developing the organization for niche GI products. Here it is already apparent that being forced into this competition given the existing structural differences is not conducive to the promotion of characteristic food products with a historical and locally embedded quality (Rosin, Stock, and Campbell 2013). The risk is that GI reinforces the hegemonic coffee value chain, including standardized and globalized forms of production, processing, distribution and even consumption.

The European Union's GI regulatory framework was established in 1992, which it modified in subsequent years in an attempt to harmonize the individual member states' legislation. The kind of protection offered by the European Union is organized in two main categories, namely PDO (protected designations of origin, identified as items "the production steps of which all take place in the defined geographical area" (EU Reg. No 1151/2012, Art. 5)) and PGI (protected geographical indications, identified as items where "at least one of the production steps of which take place in the defined geographical area" (EU Reg. No 1151/2012, Art. 5)). This now includes more than five hundred items in each category, with wine and spirits being separately regulated at the global level by the agreement of the World Trade Organization (WTO) on Trade-Related Aspects of Intellectual Property Rights (TRIPS). The European Council's database of origin and registration (DOOR) includes a huge range of food and agricultural products, as well as wines, spirits, drinks and aromatized wines.

Indonesia so far exclusively uses the PGI protection that is defined by using a name and symbol for products which were grown or processed in a particular geographical area. Although it has a long history of usage as a means of intellectual property law, with the first European laws as dating back to the 19th century, Indonesia initiated its GI law in 2007. In 2008, the first GI product registered to the Ministry of Law and Human Rights was Kintamani Arabica coffee from Bali. The growth has been slow in the years immediately after the law was introduced, but it increased dramatically starting around 2014. By 2015, there were thirty-one GI products registered, and as of 2017, thirty-seven products. The products vary from agri-food (fresh or processed) to non-food products, such as handicrafts and sculptures. Coffee did play a role in this increase, as the growth of specialty coffee in Indonesia within the last decade has been extraordinary, both among producers and consumers (Neilson 2008). By 2015, there were already eight varieties of GI coffee registered (25 percent of the total GI products), with the number increasing to twelve by 2017. This means that in 2017, a third of the registered GI products consisted of coffee. This growth (see Table 4.1) consequently shows the list of registered GI products as per 2017 data in the Ministry of Law and Human Rights. Indeed, coffee has an advantage in the sense that the unique properties of coffee can be systematically defined through the internationally accepted coffee quality spectrum, and its quality can be consistently reproduced within a particular geographical area.

Provenance for whom?

In comparing the European Union and Indonesia, we follow the tripartite understanding of GI, from earlier in this chapter, at different moments of the GI implementation: the pre-GI period, in which petty production is not unified

Table 4.1 List of GI products in Indonesia in 2017

Coffee and tea	***Spices and condiments***	***Non-agricultural food products***
Kopi Robusta Semendo	Pala (nutmeg) Siau	Garam (salt) Amed Bali
Kopi Liberika Tungkai	Lada putih (white pepper) Muntok	***Non-food agricultural products***
Kopi Arabika Simalungun	Cengkeh (clove) Minahasa	Nilam oil Aceh
Kopi Sindoro Sumbing	Vanilli (vanilla) Alor	Tembakau (tobacco) Mole Sumedang
Kopi Java Ijen-Raung	Madu Hutan (forest honey) Sumbawa	Tembakau (tobacco) Srinthil
Kopi Robusta Lampung	Mete (cashew) Bali	Tembakau hitam (black tobacco) Sumedang
Kopi Arabika Kalosi	***Staple***	Purwoceng Dieng
Kopi Arabika Toraja	Ubi (sweet potato) Cilembu	***Non-food-cultural products***
Kopi Kintamani Bali	Beras (rice) pandanwangi Cianjur	Mebel Ukir (Carved furniture) Jepara
Kopi Gayo Aceh	Padi (rice) Adan Krayan	***External products***
Kopi Bajawa Flores	***Other types of food***	Parmigiano Reggiano
Kopi Java Preanger	Kangkung (water ipomoea) Lombok	Pisco, Peru
Teh (tea) Java Preanger	Gula kelapa (palm sugar) Kulonprogo	Champagne, France
Fruits	Bandeng asap (smoked milkfish) Sidoarjo	
Jeruk keprok (mandarin) Gayo	Susu kuda liar (horse milk) Sumbawa	
Salak Pondoh		
Carica Dieng		

Source: Authors' elaboration of 2017 data of the Indonesian Directorate General for Intellectual Property Rights (DJHKI).

under a certification; at the moment of regional or global market integration; during the institutionalization process; and, finally, looking at the outcomes of GI implementation.

(1) Pre-GI: sociocultural and ecological embeddedness

The European Union's GI framework is based on the notion of "terroir" (Sylvander and Barham 2011), understood as the combined expression of place, product and people. As a core ideal of sui generis GI (European GI), it is natural to view the result of a concerted effort of human beings to understand the suitability of a land for certain products as ecological embeddedness. In the European Union, this is easy to see in documented historical terms, especially for many countries of the Mediterranean area (Sylvander and Barham 2011), but not all EU countries broadly rely on GIs for agri-food promotion and protection, which suggests that the notion of terroir is not universal. To understand how this idea is taken up in Indonesia, we want to highlight particularly the case of Ubi Cilembu (*Ipomoea batatas*), a cultivar of sweet potato grown in a village called Cilembu, just east of West Java's capital city of Bandung. Sweet potato itself is native to Latin America, and it was estimated to be introduced to Indonesia during the Portuguese era in the 16th century. As grown in different parts of Indonesia, the sweet potato developed different taste structures in the different areas. The one grown in Cilembu has a dramatically sweeter taste that appeals to popular interest, mainly in Java. Ubi Cilembu informally became a product of origin decades ago, although it was not registered as such until recently, partly because its distinctiveness is considered to highly depend on the unique ecological assemblages in Cilembu. The type of soil, weather and microbiota – or terroir in its narrowest meaning – creates such a distinct sweet potato that there has been little worry that it could be adequately grown elsewhere.

Importantly, although a concept for agri-food products of origin existed in Indonesia, divergent institutional conditions and power relations appear there only when these products must be globally recognized. In the European context, well before the European Union was established, the valorization of food items, of both their identity sense and their economic value, was part of the common vocabulary and legal understanding (Lewin 2009). In Indonesia, this is not historically true, probably due to the influence that colonizers (Geertz 1968; Booth 1989), mainly the Dutch, had exerted since the 15th century on Indonesian agriculture, which created such enduring economic relations as the spice trade and plantation commodities like coffee, tea, sugar, rubber and/or tobacco. By imposing the "*cultuurstelsel*" (or "cultivation system," a centralized agricultural policy and tax system) to maintain colonial power, the Dutch prevented the development of a system of global production and trade controlled or at least enacted by locals. Instead, vernacular forms of quality and reputability developed, for example on ways of processing coffee and tea, as in the case of *Kopi Kampong* and *Kopi Luwak*, or for trade negotiations, as in the case of *Ngadu Bako*, literally tobacco fighting, a term used in Sundanese to describe ways of tasting others' tobacco while discussing other matters. Elevating such vernacular understandings of quality to global standards, such as GI, is therefore not expected to be a linear process.

(2) (Regional/global) market integration

It is during the processes of market integration, especially at a global level, that it becomes clear how trade relations are both shaped by power relations established in colonial times or responding to new necessary logics imposed by new trade or political relations. As suggested by, among others, McMichael (2005) and Campbell (2012), the corporate model is the operative arm of neoliberal regimes, which has effectively subsumed previously vanguard alternatives (e.g., organics), and it is now turning its attention to other quality labels and to provenance. The first step here is to create a more institutionalized and legally regulated GI product, as will become clear in the next subsection. As shown in Blancaneaux (this volume), in the case of the French wine sector, the arrival of large players attracted by potential significant returns transformed GI wine from an exclusive product to a common household one. In the process, famous regions, such as Bordeaux, have undergone industrialization (Lewin 2009). Small farmers who previously controlled land and production practices and had been able to erect a quality system based on consistent results are displaced and joined by corporations, with the land price increasing substantially and the profile of wines skewed toward those that gave advantage to the new land owners (Lewin 2009).

In the case of former colonies, such as Indonesia, the culture of trader exploitation and appropriation is a long-rooted aspect of agri-food relations. For instance, the *pandanwangi* (fragrant) rice producers, who longed to reach supermarket shelves next to premium rice, ultimately found that the distribution of revenue was not nearly as fair as expected (Dwiartama, Rosin and Campbell 2016). In our case of *cilembu*, this sweet potato is still sold locally, and tourists visit the place looking for it, but producers have since realized that in Japan *cilembu* is sold for as much as US$6 a piece, of which no premium reaches the farmers or at least the region. Quality conventions are also demonstrative of exploitation, with rules for processing and control largely developed for preexisting dominant players. For instance, in the coffee sector, power relations in quality convention evolved from the long-time control by the Dutch East Indies, Chinese traders in Medan (North Sumatra), modern transnational corporations and institutions such as the Coffee Research Centre that creates cupping certificates with scoring that responds to Western consumers' tastes and expectations.

(3) GI institutionalization

The moment of institutionalization of GI, which is currently a strident neoliberal regime, also poses serious questions as to the capacity of such a measure to escape its structural constraints. The main concern here is inclusivity, or whether rightful producers are finally assigned to the area of designation and whether they can they benefit from it. But there is also a concern about how circumscribed the room to maneuver is for newcomer countries, given the technical realities of implementation in an EU-dominated field. The European Union's GI certification mechanism has a group-based application: This highlights the territorial importance of the product and the acknowledged importance of human intervention understood in a communitarian way (few exceptions are admitted). A concerted effort is necessary to obtain legal recognition, which involves both rights and duties which are often onerous for less administratively adept producers (in some cases, illiterate farmers). For example, groups must

> ensure that the quality, reputation and authenticity of their products are guaranteed on the market; take action to ensure adequate legal protection; develop information and promotion activities; develop activities related to ensuring compliance of a product with its specification; take action to improve the performance of the scheme; take measures to enhance the

> value of products and, where necessary, take steps to prevent or counter any measures which are, or risk being, detrimental to the image of those products.
>
> (Kizos et al. 2017, 2869)

All this requires a lot of time and work, which is often viable in bureaucratic systems and among producers/processors with a strong structural organization. In the EU system, national governments are responsible for organizing the mechanisms and institutions to which the aforementioned producers must refer to for proposal, implementation and maintenance of a GI. In theory, applicant producers can be small or big and yet still readily access the GI system. But even here, the system is not immune from problems of power balance. For instance, the Aceto Balsamico Tradizionale di Modena underwent a series of crises due to internal fighting about a series of fraud scandals, with large industrial producers and small artisanal ones accusing each other of not respecting the rules and pursuing different interests. The resolution came in the form of creating two different consortia aiming for wide commercial production, and the other offering a niche and expensive product. This sort of case complicates even more our understanding of GI in that it repeats long-standing issues about power and quality. But, behind this fight are highly motivated and capable parties able to express their visions and institutionalize them – a situation that cannot be taken for granted among many producer areas in Indonesia.

Another challenge in poorly resourced producer areas is that ensuring provenance for GI applications often requires historical sources. While many places in the European Union enjoy well-documented food and agriculture traditions, sometimes the documentation is so careful that the production area can be narrowly circumscribed. This does not, however, mean that slippage does not occur over time; indeed, in many cases, for instance in central Italy (Chianti Classico) and in France (Champagne), production acreage has increased. Without justification from soil or climatic conditions, such activities can be understood as merely politically or economically motivated (Lewin 2009). Capacity for documentation and traceability are also important for GIs, in which processing relies on other certified ingredients. This is, for example, the case of the origins of the pigs employed for curing meat for Parma Ham or for procuring meat from certified suppliers for the production of Aceto Balsamico Tradizionale di Modena.

In the case of Indonesian GI, the succession of laws, decrees and treaties that underpins the new mechanism has outpaced the capacity of many agencies and producer associations to understand and meet the requirements. The government decreed a law on the protection of brand and trademark only in 2001 (*Undang-Undang* No. 15/2001) that explicitly states the existence of GIs as part of the nomenclature in the context of intellectual property law. This seemed to coincide with, or be the government's complementary act in, ratifying the TRIPS Agreement of the WTO Doha negotiation round in the same year. A derivative of the law in the form of a governmental regulation

(*Peraturan Pemerintah*) was released in 2007, specifically on GIs (*Peraturan Pemerintah* No. 51/2007), which was followed by a memorandum of understanding between three ministries in Indonesia, namely the Ministry of Law and Human Rights, the Ministry of Internal Affairs and the Ministry of Agriculture, to create a legal and technical framework for the registration of GIs in the country.

Within Indonesia's regulatory framework, GI is defined as "a mark that indicates the geographical origin of a product, which due to its geographical factors, including the human and ecological factors, as well as a combination of both, provides a specific characteristic and quality to the given product" (DJHKI 2017). There are two components of a GI: its expression in letters and its denotation with a symbol. The name and the symbol should be clearly stated in the proposal when registering a product to the general list of geographical indications within the Ministry of Law and Human Rights. The products could be agriculture-, forestry- and/or fishery-based food products (fresh or processed) or non-food products such as handicrafts and sculptures. They must also comply with a certain set of requirements.

Part of registering a product as a GI includes the Ministry of Law and Human Rights (2014) asking for a requirement book (*Buku Persyaratan*), which is equivalent to the code of practice in the European GI system. It is a document consisting of information on the unique quality and characteristics that differentiate it from comparable products with the same category (e.g., coffee). The book explains in detail the name of the GI product, its physical characteristics, its ecological conditions, geographical boundaries, the history of its production/usage within the particular locale, the production process, the quality assessment method (to ensure that the quality is true and consistent) and a recommendation from authorized institutions. Table 4.2 details the components of the book.

Governmental Regulation No. 51/2007 also stipulates four conditions that prohibit a product from being registered as a GI. These conditions relate to the product's conformity to Indonesia's laws, regulations and societal values and morals. For example, alcoholic drinks cannot be registered as GIs, because the production and consumption of such products contradicts one of the religious norms in Indonesia, namely Islam. This aspect, of course, is subject to criticism and politicization. The second condition is the presentation of information that may mislead the consumers in regard to the quality, source, process and use of the product. The third condition is the use of an existing plant variety name. For instance, the *sumenep* shallot indicates a particular registered variety of shallot from the region of Sumenep, but it has long since been grown elsewhere in Indonesia. Lastly, the product cannot be registered if it has become a generic product.

It is not only the product that is subject to the law. The institution that registers this product also must fulfill particular conditions. In principle, the registrant should be an entity that acts as a representative of the product. This includes (and is limited to) gatherers (as in the case of natural products such as forest honey and wild horse milk), farmers, artisans, traders of the specific product, consumers groups or a particular institution assigned to the task of

Table 4.2 Requirements for registering a GI product in Indonesia

1	Characteristic description relates with the physical uniqueness of the GI product, which includes color, shape, texture, nutrient composition, smell, taste, etc.
2	Geographical factors include ecological factors (average temperature, rainfall, humidity, light intensity, altitude, type and texture of the soil, etc.) and sociocultural factors (traditional knowledge, social system and local economies).
3	Traditional and historical narratives covering the common practice of the local community in regard to the production and/or processing of the GI product, including the history of the product.
4	Geographical boundaries explain the delineation of the geographical area, which does not necessarily align with the administrative limits of the particular region.
5	The production process includes the steps and conditions necessary to produce a product with the given quality, which becomes a standardized production procedure for the GI product.
6	The quality assessment method involves methods for scientific assessment, such as through organoleptic checks, laboratory analysis, etc.
7	The recommendation involves a letter of support from the respective local government.

Source: Authors' elaboration of data from the 2017 Indonesian Directorate General for Intellectual Property Rights (DJHKI).

representing the product. It is recommended that the owner of the GI be a formal, multi-stakeholder institution that is able to show an effective management system, maintain the quality and consistency of the product, possess a strong marketing system, support the market demand in a sustainable manner and comply with the given regulations. Of course, such effective, organized and unified producer organizations are not a given for many well-known production areas in Indonesia.

(4) Post-GI institutionalization relations

By the time a GI is registered, if it ever reaches this stage, many criticisms related to GIs will have already become evident, and these can confirm much of the limits inherent in the GI system, but we also have found some positive unintended consequences which might open up spaces for change. As GI is structured and run, it does not provide a clear exit from structured power relations or a strong resistance to appropriation by the neoliberal regime. In the European Union, accumulated power is maintained at many levels. There is a huge debate at every announced reform of EU Common Agricultural Policy, and GI often plays a central role. In addition, we acknowledged some cases of collapse, with GIs being dismantled for failing to meet all members' requirements, solve the imbalances created in a group and even simply maintain the groups that are supposed to protect a GI product (the case of Mostviertel Perry in Austria; cf. Kizos et al. 2017). Despite Indonesia's short stint in GI, such cases are already cropping up there. These failures are substantiated by Neilson et al.'s (2018) findings about Bajawa coffee and Kintamani coffee. Likewise, from our study, in the cases of tobacco and the Java Preanger coffee, the MPIGs are non-operative, mainly due to unclear institutional arrangements and a legal framework in the protection of GI products. In Gayo Aceh (the birthplace of organic coffee in Indonesia and the only registered GI coffee in Europe), the GI functions but has not achieved significant benefits for farmers (as compared to organic and fair-trade coffee), because in practice, traders/exporters (particularly in Medan) absorb the GI premium without acknowledging and compensating for the hard work of small farmers. Generally, the coffee trade in Indonesia is still tightly controlled by former colonial powers and traders, for whom provenance is confirmed using different mechanisms (i.e., for estate-grown coffee). Perhaps most worrying but least surprising are criticisms that the legal protection for GI products do not really function better than previous informal mechanisms (Neilson, Wright and Aklimawati 2018).

Even as the technical implementation and procurement of benefits can be called into question, some positives dynamics have emerged out of the GI implementation in Indonesia. For one, the embeddedness of relationships (e.g., Morris and Kirwan 2011) has opened up space for rethinking GI, its role and its potential. Indeed, in Indonesia, some coffee farmers lament that local middle-class consumers do not understand the legal consequences of GIs but value the certification regardless, as a symbol and as an identity of their products. In the European Union, although GIs are ubiquitous, some wine producers have gathered together to renounce to the highest legal protections, to pursue ideals of production and community that may be incompatible with GI or other market options (Castagno, Gravina and Rizzari 2018).

Conclusion

GI has been depicted as a new frontier for creating extra value for global producers of distinctive agri-food products and beverages, although the accumulated literature continues to highlight the limitation of the expansion of GI. The two cases we present here, the European Union and Indonesia, differ enormously because of different historical, institutional, organizational and underlying ecological characteristics, but many of the unfolding lessons are similar. In both cases, there are unceasing negotiations about gaining or obtaining legal recognition for food products that raise questions about the potential for GI to carve a fairer niche in global commerce. In both regions, there are challenges of inclusivity, in which advantaged and powerful players become gatekeepers, with a high level of motivation to defend rentier power. And lastly, both regions struggle to reconcile the useful fluidity of the concept of terroir (recognition of human, environment and historical contribution) with the instrumentalization of the concept for inclusion or exclusion. Given these ongoing struggles, combined with the administrative capacity necessary to effectively navigate the complex GI institutionalization, is it reasonable to expect that newcomer GI countries will be able to design systems that escape neoliberal market dominance? Can the sociocultural embeddedness in the practice of GI be saved through innovation? As it stands, GI is not trending as a strong alternative to the neoliberal regime, as Quiñones-Ruiz et al. (2015) advocate, and may be introducing new challenges for producers of distinctive food products. Given what happened in the European Union, it is important to anticipate which pathways may be experienced in developing countries like Indonesia and to preemptively act on this knowledge. The potential for GI to revitalize sociocultural relations and encourage more vibrant local economic relations remains a hopeful ideal, but reaching this ideal requires reflection and empowerment that has mostly not developed yet. McMichael (2013) found hope in the disrupting potential of peasant groups who are consciously organized and willing to reach their goals; whether this can hold true for GI is likely also a matter of concerted efforts. For whom provenance is an advantage should remain a dogged question, since the risk of perpetuating structural disadvantages; otherwise, historically submissive relations will remain hidden in the very branding of provenance as a new tool for development.

Bibliography

Augustin-Jean, Louis. 2012. "Introduction: The Globalization of Geographical Indications: The Challenge for Asia." In *Geographical Indications and International Agricultural Trade: The Challenge for Asia*, edited by L. Augustin-Jean, H. Ilbert and N. Saavedra-Rivano, 1–16. London: Palgrave Macmillan.

Booth, Anne. 1989. "Indonesian Agricultural Development in Comparative Perspective." *World Development* 17 (8): 1235–54.

Bowen, Sarah. 2010. "Embedding Local Places in Global Spaces: Geographical Indications as a Territorial Development Strategy." *Rural Sociology* 75 (2): 209–43.

Busch, Lawrence. 2011. "Food Standards: The Cacophony of Governance." *Journal of Experimental Botany* 62 (10): 3247–50.

Campbell, Hugh. 2012. "Let Us Eat Cake? Historically Reframing the Problem of World Hunger and Its Purported Solutions." In *Food Systems Failure, the Global Food Crisis and the Future of Agriculture*, edited by Chris Rosin, Paul Stock and Hugh Campbell, 30–34. London, UK: Earthscan.

Carolan, Michael. 2015. "Affective Sustainable Landscapes and Care Ecologies: Getting a Real Feel for Alternative Food Communities." *Sustainability Science* 10 (2): 317–29.

———. 2013a. *Embodied Food Politics*. London, UK: Routledge.

———. 2013b. "Final Word: Putting the" Alter" in Alternative Food Futures." *New Zealand Sociology* 28 (4): 145–50.

Castagno, Armando, Giampaolo Gravina, and Fabio Rizzari. 2018. *Vini Artigianali Italiani*. Roma, Italia: Paolo Bartolomeo Buongiorno Editore.

Dwiartama, Angga, Chris Rosin, and Hugh Campbell. 2016. Understanding Agri-food Systems as Assemblages." In *Biological Economies: Experimentation and the Politics of Agri-food Frontiers*, edited by Richard Le Heron, Hugh Campbell, Nick Lewis and Michael Carolan, 82–94. London: Routledge Publishing.

Friedmann, Harriet. 2016. "Commentary: Food Regime Analysis and Agrarian Questions: Widening the Conversation." *The Journal of Peasant Studies* 43 (3): 671–92.

Geertz, Clifford. 1968. *Agricultural Involution: The Process of Ecological Change in Indonesia*. No. 11. Los Angeles, CA: University of California Press.

Granovetter, Mark. 1985. "Economic Action and Social Structure: The Problem of Embeddedness." *American Journal of Sociology* 91: 481–510.

Henry, Matt, and Michael Roche. 2013. "Valuing Lively Materialities: Bio-economic Assembling in the Making of New Meat Futures." *New Zealand Geographer* 69 (3): 197–207.

Hughes, Justin. 2017. "The Limited Promise of Geographical Indications for Farmers in Developing Countries." In *Geographical Indications at the Crossroads of Trade, Development, and Culture: A Focus on Asia-Pacific*, edited by I. Calboli and N. L. W. Loon, 61–86. Cambridge: Cambridge University Press.

Indonesia's Directorate General for Intellectual Property Rights (DJHKI). 2017. "Pengenalan Indikasi Geografis." Accessed on August 1, 2017. www.dgip.go.id/pengenalan-indikasi-geografis.

Kizos, Thanasis, Ryo Koshaka, Marianne Penker, Cinzia Piatti, Christian Reinhard Vogl, and Yuta Uchiyama. 2017. "The Governance of Geographical Indications: Experiences of Practical Implementation of Selected Case Studies in Austria, Italy, Greece and Japan." *British Food Journal* 119 (12): 2863–79.

Lewin, Benjamin. 2009. *What Price Bordeaux?* Paris: Vendange Press.

Loconto, Alison, and Lawrence Busch. 2010. "Standards, Techno-economic Networks, and Playing Fields: Performing the Global Market Economy." *Review of International Political Economy* 17: 507–36.

McMichael, Philip. 2013. *Food Regimes and Agrarian Questions*. Halifax: Fernwood Publishing.

———. 2005. "Global Development and the Corporate Food Regime." In *New Directions in the Sociology of Global Development (Research in Rural Sociology and Development, Volume 11)*, edited by Frederick H. Buttel and Philip McMichael, 265–99. Emerald Insight: Emerald Group Publishing Limited.

Morris, Carolyn, and James Kirwan. 2011. "Ecological Embeddedness: An Interrogation and Refinement of the Concept Within the Context of Alternative Food Networks in the UK." *Journal of Rural Studies* 27 (3): 322–30.

Murray Li, Tania. 2007. "Practices of Assemblage and Community Forest Management." *Economics and Society* 36: 263–93.

Neilson, Jeffrey. 2008. "Global Private Regulation and Value-chain Restructuring in Indonesian Smallholder Coffee Systems." *World Development* 36 (9): 1607–22.

Neilson, Jeffrey, Josephine Wright, and Lya Aklimawati. 2018. "Geographical Indications and Value Capture in the Indonesia Coffee Sector." *Journal of Rural Studies* 59: 35–48.

Quiñones-Ruiz, Xiomara F., Marianne Penker, Christian R. Vogl, and Luis F. Samper-Gartner. 2015. "Can origin Labels Re-shape Relationships Along International Supply Chains? – The Case of Café de Colombia." *International Journal of the Commons* 9 (1): 416–39.

Rosin, Christopher, Paul Stock, and Hugh Campbell, eds. 2013. *Food Systems Failure: The Global Food Crisis and the Future of Agriculture*. London: Routledge Publishing.

Sylvander, Bertil, and Barham Elisabeth. 2011. "Introduction." In *Labels of Origin for Food: Local Development, Global Recognition*, edited by Bertil Sylvander and Elisabeth Barham, ix–xviii. Wallingford, UK: CABI.

Tansey, Geoff, and Tony Worsley. 1995. *The Food System: A Guide*. London: Earthscan Publications Ltd.

Wald, Navé, and Douglas P. Hill. 2016. "'Rescaling' Alternative Food Systems: From Food Security to Food Sovereignty." *Agriculture and Human Values* 33 (1): 203–13.

Part III

Cases from Europe

5 How to use geographical indication for the democratization of agricultural production

A comparative analysis of geographical indication rent-seeking strategies in Turkey

Derya Nizam

Introduction

Geographical indication (GI) distinguishes selected products from other products that have similar characteristics and are produced in different cultural and ecological environments. Rural producers and communities have been increasingly using GIs to differentiate their local products from anonymous mass-produced commodities (Barham 2003). As defined by David Harvey (2001, 395), monopoly rents emerge when there is exclusive control over a tradable item that is unique and non-replicable in certain crucial aspects. In the case of the GI label, rent may be viewed as based on access to a specific terrain that is associated with distinctive characteristics. Since land is a scarce means of production, unequal access to this certified (limited) land works as a formidable barrier to entry (Guthman 2007).

However, creating rent from GI protection depends not only on the scarcity of certified land but also on cultural practices maintained for the social reproduction of a certain uniqueness of place (Coombe and Aylwin 2011). The perceived distinctive quality is also associated with cultural practices, historical identities and ecological or biological resources that have characterized a region for generations (Bowen 2010). From this perspective, GI protection has two elements of rent formation. The first refers to claims about a special territorial quality. The second consists of limiting the conditions of reproduction of this territorial quality through the GI governance system.

This study identifies and illustrates the conditions for such rent-seeking capacities through GI systems by focusing on rent creation (Nizam 2017a). In a GI protection setting, rent creation refers to the characteristics of GI processes that constitute barriers to entry for competitors that local actors use to distinguish their products (such as limiting the production area; specifying a certain variety; or adopting codes of practices, verification and control) (Nizam 2017a). In this study, rent creation is analyzed to identify the special characteristics of GI processes, which in turn align with many governance structures in product supply chains. In a similar way, Niederle, Wilkinson and Mascerenhas (this volume) introduce the term *project effect* to refer to the effects of governance mechanisms of GI projects. Table 5.1 shows key characteristics of products that can be barriers to GI supply chains based on *quality*, *geography* and *control*. The juxtaposition of quality, geography and control creates a layout for a comparison that explains the varying effects of the three Aegean olive oil GIs (*Edremit Gulf Region Olive Oil GI 2004, South Aegean Olive Oil GI 2004 and Ayvalik Olive Oil GI 2007*).

In Turkey, GI protection was first put into practice in 1995 to provide a legal framework for the protection of the names of products generated with resources and/or methods exclusively attached to a region. The European Customs Union membership process that began in 1995 obliged Turkey to introduce the basic jurisdiction of intellectual property rights (European Commission 2006). The 1995 Law No. 555 regulates the following two types of GIs: PDO and PGI. The definitions of these types correspond to the European Economic Community (EEC) Regulation No. 2081/92.

This legislation was strengthened by the law of 6 October 2003, which led to the creation of the Turkish Patent Institute (TPI). Protection of GIs, as well as patents, trademarks, designs and integrated layout designs, are within the competence of TPI (Ozbag 2017). The Law on Industrial Property No. 6769 (hereafter, in this chapter, "the Law") which was adopted on 22 December 2016 by the Turkish Parliament changed the name of the TPI to the Turkish Patent and Trademark Office (TPTO). In Turkey, by September 2018, 375 GI products have been certified, and 403 GI applications were pending.

In Turkey, GI is still new, accordingly the implications of GI projects in the generation of inclusionary social consequences (e.g., protecting biodiversity, promoting rural development and ensuring the fair distribution of value added) cannot be assumed, given that local producers are relatively unfamiliar with GI protection (Nizam 2017b). In general, in Turkey, GIs have not worked effectively in generating economic benefits for producers. Yet, some research shows

Table 5.1 The role of policy rents in the design process of three olive oil GIs

GIs	*Edremit Gulf Region Olive Oil*	*South Aegean Olive Oil*	*Ayvalik Olive Oil*
Quality (product under GI protection)	Homogenous/mass quality	Homogenous/mass quality	Heterogonous /niche quality
Geography (volume of production under GI protection)	Medium volume	High volume	Small volume
Control (participation and governance in decision-making process of GI)	Top-down approach Vertical coordination	Top-down approach Vertical coordination	Bottom-up approach Horizontal coordination

Source: Nizam 2015.

that in some instances, GI policies have been effective in achieving local market stability against the detrimental effects of trade liberalization (Nizam 2009). Because the impacts of GI are thus difficult to benchmark, this study does not present a cost-benefit analysis of data on GI protection but rather presents an investigation of the value for rural livelihoods as perceived by local actors.

The number of participants using a specific GI is interpreted as a matter of effectiveness, which reflects a degree of internal coordination and collective action among different producers. Each GI is taken as a unit of analysis and compared through a disarticulation approach to global commodity chains. This comparison focuses on the processes of "articulation" (linking) of people and places to particular chains and "disarticulation" (delinking) of them from others (Bair and Werner 2011). The idea here is to challenge the inclusionary bias of commodity chain analysis and shift the focus of analysis to the various kinds of exclusions.

Reflexive politics of the local

The goal of this study is to analyze the potential for GIs to create, capture and distribute rent in ways that promote a reflexive localism that empowers rural actors and communities through alternative schemes. These are schemes that aim at reducing stakeholders' dependence on corporate agri-food. Instead of naturalizing and fetishizing the local as intrinsically just in specific ways, we should critically think about how to make local food systems more just in a reflexive politics (DuPuis and Goodman 2005, 364). Otherwise, GI systems turn into "a way for local elites to create protective territories for themselves" (DuPuis and Goodman 2005, 364). This eventuality would fundamentally undermine the ideals of GI. For instance, Renard and Domínguez Arista (this volume) explain that the introduction of GI protection to mezcal production deepened the power concentration process in which traditional artisanal producers of mezcal were excluded by the industrial producers. In a similar vein, Feuer (this volume) shows that the GI protection for Kampot pepper in Cambodia contributed to the exclusion of local producers from the supply chain of the crops and to the control by predominantly foreign intermediaries in the value chain.

Understanding which of the options local actors negotiate in the design and maintenance of production conditions in a GI protection setting reveals the significance of rent creation in two ways. The first is the effectiveness of the valorization process, and the second is its redistributive potential (DuPuis and Goodman 2005). There are many GIs that do not result in economic premiums. Additionally, in the event that these premiums are achieved, the very process of negotiating the norms that regulate the GI process may lead to hidden powerful supply-chain actors capturing these premiums (Bowen 2010). GIs that privilege industrial standards may turn into "relic spatial strategies" for industrial producers, whereby even remote rural areas become subordinate to the global economic system and its decision-making pathway (Wilson and Whitehead 2012, 205).

In short, the notion of reflexive localism provides a critical perspective to trace a number of strategies used in the process of designing GI systems, and questions their potential to empower rural actors and communities by creating opportunities for extracting economic rent. The potential for GIs to empower rural producers in global markets is contingent upon the particular local governance mechanisms in place (Bowen and De Master 2011; Neilson 2007; Pacciani et al. 2001; Paus and Reviron 2010).

Research setting and methodology

Fieldwork centered on three olive oil GIs in the Aegean region, namely in the *Edremit Gulf Region*, *South Aegean* and *Ayvalik*. These products were selected as typical Aegean products because they are associated with the geographic region and have a well-established reputation. However, there are important differences between their associated commodity chains in terms of natural and cultural ecologies and in how people are positioned within, and related to, the chains.

The Aegean region of Turkey has historically been able to achieve high prices for olives, its leading agricultural crop. However, since 2000, the region has risked losing its leadership status due to the high production costs tied to rugged land characteristics (Nizam 2017b). This became the case with the implementation of neoliberal policies whereby olives became an alternative crop for farmers involved in industrial agriculture. In this context, labor costs in olive production declined as large farmers adopted mechanical harvesting techniques that were possible because of the suitable land characteristics in the new plantations. Simultaneously, the high production costs within the Aegean region and the quality and reputation of the local olives allowed for the emergence of GI protection that translated into the preferred alternative economic strategy in its competition with other regions.

Edremit Gulf Region olive oil and South Aegean olive oil are two GIs designed by the executives of Taris, the sales cooperative representing these GI areas. As an olive and olive oil agricultural sales cooperative, Taris plays a leading role in the region and in the domestic market. Since their registration in 2004, five firms (including Taris) have made use of the Edremit Gulf olive oil GI, and two firms (including Taris) have made use of the South Aegean olive oil GI. The effectiveness of these two GIs is, however, questionable.

Shortly after the registration of Edremit Gulf Region GI, Taris came into conflict with one of the cities that is included within the borders drawn for this GI production area. The city of Ayvalik immediately applied to the TPI for a new GI for production carried out within its city borders. After a long dispute between the parties, a GI application for Ayvalik olive oil sponsored by a local initiative led by the Ayvalik Chamber of Commerce (ACC) was approved in 2006. Since this date, the olive oil produced within the Ayvalik city limits can be legally marketed by both GI labels, but the Ayvalik GI is considered more effective than other GI because forty-six firms have regularly made use of the

Ayvalik GI localism. As mentioned, in this study, the number of participants using a specific GI is taken as a matter of effectiveness that reflects its attractiveness for rural actors and its impact on rural society in organizing collective action that symbolizes community resilience. In other words, this study assumes that a project is effective when it increases the ability of rural actors and communities to organize collective action. The number of participants using GI is taken as an indicator of how effective these GI projects are in achieving their goals.

In the initial fieldwork conducted in the fall of 2011, sixteen in-depth interviews were conducted with producers and nine key informant interviews with professionals. A second phase conducted in the summer of 2012 consisted of fifty-six in-depth interviews and 150 survey interviews with producers in the same area. A follow-up phase was conducted five years later. Secondary sources, such as official agricultural production reports, trade statistics and newspaper articles, were reviewed for the analysis of differences among commodity chains concerning natural and geographical properties, labor relations, industrial upgrading and local know-how.

The role of rent creation in the geographical indication design process

The varying effects of GI protection are identified through a comparative analysis of the options negotiated by local actors in the design of the three different Aegean olive oil GIs. The differentiation of rent creation across the three olive oil GIs is summarized in Table 5.1, which shows that rent creation depends on three factors. The first factor is the capacity of the GI to separate the product and its characteristics from other products by establishing an external reputation based on the link between the product and its origin. The second factor is the limiting of the geographical area under GI protection, such that the volume of production shows that the link between the distinctive quality and the origins is not arbitrary. The last factor is the participation and governance constructed and maintained both in the design process and in the verification systems. These three links are explored further below.

Product characteristics under geographical indication protection

GI protection is based on monopoly formation, which is based on the establishment of barriers to entry. Specific definitions regarding who can make a GI product, where and through what techniques set these barriers to entry (Rangnekar 2011, 2047). Such definitions are reflected in GI descriptors. In all three Aegean olive oil GIs, the variety of the olive grown was determined by the GI designers to be one of the defining identities of the GI product. In comparison, the designers paid less attention to the historical elements, traditional practices and know-how of local farmers.

The local variety (Ayvalik – Edremit Yaglik) is seen as important for rent creation since it differs from another domestic variety, the Gemlik variety, which has been widely planted as a monoculture in other regions of Turkey. The Gemlik variety has a strong reputation among table olives and can be produced in a short time period in its original agri-ecosystem. However, it provides low-quality fruit in plantations out of its home ecosystem, and accordingly, it is often used for cheaper oil instead of table olive production. This means that regions adopting this variety are de facto turned into cheap oil producers, increasing regional commodity competition. The region lost its chance to compete with other regions based on price, due to the high production costs associated to the hilly and mountainous characteristics of the area (Nizam 2017b). Local actors try to improve competitiveness by selling their olive oil with a territorial quality, which consumers are willing to pay.

Defining territorial quality become a crucial site of power relations through which meanings of territorial quality and associated practices are negotiated, contested and eventually reified. A plurality of agents compete for the authority to define the main characteristics of territorial quality. The scientism discourse (the belief that policy can be best dictated by science) encourages local actors to assume an "ontological separation" between human and nonhuman factors (Goodman 2001) and then defend nonhuman factors (such as soil, climate and winds) as neutral, objective and reliable indicators of "an ideal GI boundary." Accordingly, local actors believe that local taste largely depends on nonhuman factors that extra-local actors cannot easily appropriate or substitute. From this point, they argue that human characteristics play less of a role in shaping quality, because they perceive these characteristics as movable effects that can be reproduced by extra-local actors. Local actors seem to expect GI protection to place limits on agribusiness in controlling natural processes. In other words, the bias for nonhuman factors is institutionalized by local actors themselves because of their perpetual search for monopoly rent that seeks out uniqueness disarticulated from the appropriation/substitution machine of agribusiness (Nizam 2015).

In contrast, in both the South Aegean GI and the Edremit Gulf Region GI, it was decided to *regularly* adjust the code of practices to meet the requirements of national and international food codes. These two GIs reflect a "relic spatial strategy" where the local became subordinate to the global economic system and its decision-making pathways (Wilson and Whitehead 2012). In practice, the GIs reflect more modern "codes of practice," which are usually designed in line with the principle of homogenous and mass quality, with the goal of meeting the requirements of retail industry. The broader aim of Turkish food codes is to create commercial standards that work as a mass quality stabilizer. This contradiction was articulated by a representative from the governing cooperative, who argued that the use of traditional techniques was risky and that the production standards identified in the food code were an essential part of preserving quality and a GI's reputation:

> Traditional methods are a very controversial issue that must be taken seriously. The possibility of producing substandard oil is very high when traditional methods are used.
>
> (A representative of the governing cooperative, Izmir, 2012)

The Ayvalik GI, in contrast, offers more flexibility in olive pressing methods by having two different quality management systems in place: one for traditional presses and the other for modern press processes. Since the sensory characteristics of Ayvalik olive oil are said to be strictly tied to the locale – where variety, microclimate and particular skills give the fruit its characteristics – local know-how is also seen as an important resource that can be protected through rent creation. Ayvalik's code of practices have been designed with heterogeneous quality in mind, and this flexibility in press methods provides all producers with a chance to enter GI chains. Ayvalik GI benefits from its reputation, developed and maintained by traditional and artisanal methods. However, a portion of its production is actually based on industrial methods. Through the label of stone mill pressed, high-cost traditional processing can be recognized and rewarded by consumers. A local producer/trader who is also a member of the ACC argues that both traditional and industrial standards are important to preserve the reputation of their oil:

> We created a space for both traditional and modern press methods. In terms of the industry, we are very lucky because the chemical values of the oil are very compatible with industrial standards. The shelf life of the Ayvalik oil is quite long. . . . This is the scientific aspect of the fact. In terms of tradition, yes, Ayvalik olive oil has a reputation derived from traditional characteristics.
>
> (A member of ACC, Ayvalik, 2011)

Strict quality requirements may reduce producers' ability to negotiate exceptions or adopt new methods according to a fair timeline (Rangnekar 2011). For instance, a regulation in the Turkish food code recently made stainless steel tanks a requirement in oil storage; however, the majority of the farmers on the Edremit Gulf will be unable to afford them. Stainless steel tanks are required for South Aegean GI and the Edremit Gulf Region GI, because they need to conform to the code of practices defined in national food codes. This is not required for Ayvalik. Producers who cannot afford the required storage tanks or meet other conditions are supposed to transfer their whole yield to industrial mills after it has been pressed, meaning they have to sell their product during high season (harvest) when prices are usually at their lowest. This will likely reinforce power asymmetries and the lopsided distribution of added value along the supply chain in favor of certain powerful actors.

Volume of production under geographical indication protection (the limitation of geographical indication areas)

Every GI protection involves the practices of setting boundaries and governing those boundaries (Guthman 2007). Although local actors set boundaries, they do not just limit the physical space but also form new cultural identities, proximity and social differences (Coombe and Aylwin 2011). The geographical area under the GI protection of South Aegean olive oil is approximately

35,500 km², Edremit Gulf olive oil 6,000 km² and Ayvalik olive oil 265 km². Galtier, Belletti and Marescotti (2008) show how the redefinition of the link between the products' territorial quality and the volume of production can serve as one of the strategic options in the design of GIs. High territorial identity might, in the case of mass production, create an origin-based differentiation attractive to niche intermediate markets. However, in small-scale production, high territorial identity targets niche final consumers. Local exporters recognize this:

> GI can protect only boutique producers in a globalized world, it cannot increase bulking (in volume) . . . to be successful in bulk markets requires the national presentation as a trademark, like in the case of Spain and Italy.
>
> (An olive oil exporter, Izmir, 2011)

Whereas South Aegean and Edremit Gulf Region GIs reflect macro regions with associated mass production that target niche intermediate markets through rent creation of flavor and taste, the Ayvalik GI reflects a micro region and small scale of production which targets niche final consumers primarily through a rent creation based on reputation. In designing GIs, the executive directors of Taris perceive a strong rent creation by arguing that the local variety is the best indicator to set GI boundaries. Based on three different, yet related, strategies, these boundaries are established as immovable and historical facts by the GI designers. The first claims an objective, tangible and conflict-free indicator of quality. The second claims an important cultural heritage that consumers will see as a collective asset. The third is based on the guarantee of large-scale production and product differentiation that allow the governing cooperative to exploit economies of scale. Boundaries naturally drawn by local varieties divide the whole Aegean region in two large segments. This situation guarantees the large-scale production of local varieties.

Targeted production volume is one strategic option in setting the boundary between quality and geography based on targeted markets (Galtier, Belletti and Marescotti 2008). Accordingly, the strategy to establish a macro region aims at enabling mass production that targets niche intermediate markets. Its clients are wholesale markets that supply supermarket chains such as Metro, and grocery stores or retailers sourcing oil for private label sales. Taris is a major brand that combines oil from thousands of producers to attain an ample and standard quality throughout the year. From their perspective, scarcity is seen as manageable at a certain scale of production, because mixing oils is unavoidable for large-volume actors like Taris.

In contrast to the Edremit and South Aegean olive oil whose GIs' boundaries are drawn by local variety, the Ayvalik GI fits the physical boundaries of the city of Ayvalik. The product characteristics, including organoleptic quality and reputation, are attributed to the landscape, climate, soil and local know-how of this physical and cultural locale. The first proposed GI boundaries were extended beyond the city limits by including land that was within the borders of Ayvalik during the Ottoman Empire period but that is currently incorporated within a nearby city. In 2007 and before making a decision on the GI status for this product, the proposed boundaries were challenged and conformed to the city limits by the TPI due to the presumption that contemporary political boundaries align with social and ecological boundaries. Some traders still criticize this decision by TPI:

> The GI area is necessarily equal to the city limits. Ayvalik olive oil has already a traditional characteristic, it hasn't been recently invented. It has had a reputation for many years, for centuries, we just certified it . . . and we try to sell it now with that story.
>
> (A representative of ACC, Ayvalik, 2011)

Ayvalik uses this reputation in establishing cultural identity and proximity to consumers through a nostalgia for the perceived multiculturalism of Ottoman times. Family stories and memories of farmers are used by firms in Ayvalik to reflect a difference in their quality coming from multicultural and hybrid cultural values. Tourism reinforces this, playing an important role in the recognition of human and nonhuman geography that can be related to the GI protection. In contrast, there is no rent creation for reputation at work with the other GIs obtained by Taris. As a small area and tourist destination, the Ayvalik GI therefore aims at capturing growing demand directly from niche final consumers.

Participation and governance in decision-making processes of geographical indication

Kaplinsky and Fitter (2004) show that rent-generating barriers create higher returns and that governance dynamics describe which firms or actors determine how those returns are distributed. Instead of individual appropriation, a plurality of actors can employ GIs as property. However, this opportunity is strictly based on the capacity of local actors to create the institutional processes necessary for controlling and regulating the conditions required to be entitled to GI labels (Pacciani et al. 2001). This means that the institutional environment is itself a rent-generating entity that transforms added value into economic rent and protects and governs territorial quality through grades, standards and certification.

From this point of view, GI is recognized and constructed as a rent generator for local resources, in terms of not only external recognition but also internal coordination. The three Aegean cases offer contrasting models for such internal coordination. The South Aegean and Edremit Gulf GIs are based on a top-down project in which executive directors used technical parameters to determine the primary codes of practices (product characteristics and volume of production). The guarantee system and its sanctions were designed and set by executive directors in a top-down model rather than by a group of rural producers in a bottom-up model. The actual producers were not encouraged to participate in the governance of these two GI projects. As Giovannucci and Ponte (2005) indicate, although some certifications may provide substantial benefits for producers, existing power relations may remain essentially unchanged or be perpetuated if producers have no control over key decision-making processes. As Sekine (in this volume) notes, GI systems do not systematically contribute to the democratization of agri-food systems and sustainable territorial development. She also stresses that public institutions and their officers may assist GI stakeholders in achieving these goals. However, the active participation of local actors in decision-making processes is vital to achieving this objective.

In the case of the Ayvalik GI, the registration of the GI was able to be achieved through strong cooperation and coordination created and implemented by local actors working under the local chamber of commerce. The registration was proposed and completed thanks to strong collective action and with the awareness and involvement of local traders in Ayvalik. Participants in the Ayvalik GI administration have explicitly formulated their approach in opposition to that of the cooperative that sponsors the South Aegean and Edremit Gulf GIs:

> Nothing changes when a GI is registered. What is important is to provide a secure supply system and to create brand awareness as a whole. Taris registered the GI for a wide area, and it is not easy for them to protect the local characteristics and control the GI chain in that wide area.
>
> (A bureaucrat from ACC, Ayvalik, 2011)

The changing involvement of local producers in the three GIs derives from differences in the code of practices, the volume of production and the control forms that generate rent. The governance of the South Aegean and Edremit Gulf GIs reflects a "normative localism" in which a small group of executive directors decides what is best for the producers and defines who and what is included and excluded (DuPuis and Goodman 2005). In this case, democratic deficit emerged even before the process of the GI's area identification was set in motion. Conversely, the governance of the Ayvalik GI represents an instance of "reflexive localism" in which representatives of local actors are brought in to deliberate on how they want to change their society. This remains the case even in the event that contradictions and conflicts of interest emerge (DuPuis and Goodman 2005). This is not only an important strategic option that is recognized and built through the GI protection but also an important disarticulation moment that separates GI from industrial chains by creating a change in the governance of supply chains.

Conclusion

This chapter provides empirical evidence of how local actors perceive GIs as effective tools to address the increased pressure of "economies of scale" and "standardization" in agricultural production in a specific area of Turkey.

Following established analyses, the chapter contends that the question of whether GIs and the accompanying efforts have a positive effect on democratic forms of development is an empirical matter (Rangnekar 2011). GI's potential impact on rural communities and agri-food can vary. And this variation depends on the ways local stakeholders act in regard to these labels (Nizam 2015). The case studies illustrate that political, cultural and ecological boundaries do not necessarily overlap and that GI protection triggers interregional or intraregional competition within a new self-governing development regime. They also show that the "rural" emerges as a site of social innovation and entrepreneurship as rural areas adapt to the changing conditions imposed by the existence of the neoliberal regime (Anthopoulou, Kaberis, and Petrou 2017).

Following the analyses presented above, GIs have arguably emerged as a tool to protect local producers and economies from the unwanted consequences of enhanced global competition (DuPuis and Goodman 2005). From this perspective, GI projects can be considered not so much as a form of resistance to the neoliberal regime but rather as an intrinsic part of it. Consequently, it may be appropriate to discuss GI projects in the context of the larger debates about the devolutionist forms of governance – that is, self-governing projects that promote corporate interests and the progressive ideals of economic development in a neoliberal age. However, this study also presents evidence that some rural actors are using GIs as a form of resilience and as a means of preserving the traditional farming and knowledge systems that are vital to their livelihood.

The certification process is a site of critical power relations, and accordingly, it becomes a center of social interaction for grassroots rent generation (Mutersbaugh 2005). This chapter concludes that the current global expansion of GIs offers an opportunity to examine emerging forms of political subjectivities. To put it more directly, by making a comparative analysis of how GI initiatives reinvent markets through a collective label, this chapter reveals the importance and the constitutive role of collective action in protecting and governing the territorial quality that is embedded in the local terroir. The appropriation of culture and history in the definition of the collective rules of production is crucial, and it arms GIs with the capacity to revitalize rural areas by giving voice to, and empowering, local actors. Existing power relations could stay essentially unchanged or be perpetuated if local actors remain on the receiving end of key decision-making processes. For this reason, the chapter concludes that GI application procedures should become a collective learning or decision-making process that is integrated into a locally directed and more comprehensive developmental strategy. The possibility of activating networked forms of coordination based on GI depends on how strong the association or link between the product and the local community is. Indeed, the ability to make rules linked to a specific context and to reproduce the uniqueness of territorial quality through traditional and cultural techniques is a prerequisite for the legal recognition of GI, and this ability represents a foundation on which local institutions can be built and policies designed that can offset the growing power of corporate agri-food.

Bibliography

Anthopoulou, T., N. Kaberis, and M. Petrou. 2017. "Aspects and Experiences of Crisis in Rural Greece: Narratives of Rural Resilience." *Journal of Rural Studies* (52): 1–11.

Bair, J., and M. Werner. 2011. "Commodity Chains and the Uneven Geographies of Global Capitalism: A Disarticulations Perspective." *Environment and Planning A* 43: 988–97.

Barham, E. 2003. "Translating Terroir: The Global Challenge of French AOC Labeling." *Journal of Rural Studies* 19: 127–38.

Bowen, S. 2010. "Embedding Local Places in Global Spaces: Geographical Indications as a Territorial Development Strategy." *Rural Sociology* 75: 209–43.

Bowen, S., and K. De Master. 2011. "New Rural Livelihoods or Museums of Production? Quality Food Initiatives in Practice." *Journal of Rural Studies* 27: 73–82.

Coombe, R. J., and N. Aylwin. 2011. "Bordering Diversity and Desire: Using Intellectual Property to Mark Place-Based Products." *Environment and Planning A* 43: 2027–42.

DuPuis, E. M., and D. Goodman. 2005. "Should We Go 'Home' to Eat? Toward a Reflexive Politics of Localism." *Journal of Rural Studies* 21: 359–71.

European Commission. 2006. "Screening Report Turkey, Chapter 11 – Agriculture and Rural Development." Online document. Accessed March 10, 2008. http://ec.europa.eu/enlargement/pdf/turkey.

Galtier, F., G. Belletti, and A. Marescotti. 2008. "Are Geological Indications a Way to 'Decommodify' the Coffee Market?" International Congress of European Association of Agricultural Economists, August 26–29, Ghent and Belgium.

Giovannucci, D., and S. Ponte. 2005. "Standards as a New Form of Social Contract? Sustainability Initiatives in the Coffee Industry." *Food Policy* 30: 284–301.

Goodman, D. 2001. "Ontology Matters: The Relational Materiality of Nature and Agri-food Studies." *Sociologia Ruralis* 41: 182–200.

Guthman, J. 2007. "The Polanyian Way? Voluntary Food Labels as Neoliberal Governance." *Antipode* 39: 456–78.

Harvey, D. 2001. "The Art of Rent: Globalization, Monopoly and the Commodification of Culture." In *Socialist Register 2002: A World of Contradictions*, edited by L. Panitch and C. Leys. London: Merlin Press.

Kaplinsky, R., and R. Fitter. 2004. "Technology and Globalisation: Who Gains When Commodities Are De-commodified?" *International Journal of Technology and Globalisation* 1: 5–28.

Kaplinsky, R., and M. Morris. 2001. "A Handbook for Value Chain Research." Working Paper Prepared for the IDRC, Institute for Development Studies, Brighton.

Mutersbaugh, T. 2005. "Fighting Standards With Standards: Harmonization, Rents, and Social Accountability in Certified Agrofood Networks." *Environment and Planning* 37: 2033–51.

Neilson, J. 2007. "Institutions, the Governance of Quality and On-farm Value Retention for Indonesian Specialty Coffee." *Singapore Journal of Tropical Geography* 28: 188–204.

Nizam, D. 2017a. "Place-based Labels in Agricultural Value Chains." *Comparative Sociology* 16 (3): 422–45.

———. 2017b. "Place, Food and Agriculture: The Use of Geographical Indications in Turkey." *New Perspectives on Turkey* (57): 3–30.

———. 2015. "Geographical Indications and Commodity Chain Analysis: Policy and Resource Rents." Unpublished PhD Thesis, The University of Sydney, Australia.

———. 2009. "Protection of Geographical Indication: A Study of 'Aegean Cotton Mark'." Unpublished Master Thesis, Bogazici University, Turkey.

Ozbag, N. 2017. "Geographical Indications in Turkey." Report by USDA Foreign Agricultural Service. Report no: TR7057.

Pacciani, A., G. Belletti, A. Marescotti, and S. Scaramuzzi. 2001. "The Role of Typical Products in Fostering Rural Development and the Effects of Regulation (EEC) 2081/92." In the 73rd Seminar of the European Association of Agricultural Economists, Ancona.

Paus, M., and S. Reviron. 2010. "Crystallisation of Collective Action in the Emergence of a Geographical Indication System." International EAAE-SYAL Seminar – Spatial Dynamics in Agri-food Systems, Parma.

Rangnekar, D. 2011. "Remaking Place: The Social Construction of a Geographical Indication for Feni." *Environment and Planning A* 43: 2043–59.

Wilson, G. A., and I. Whitehead. 2012. "Local Rural Product as a 'Relic' Spatial Strategy in Globalised Rural Spaces: Evidence from County Clare (Ireland)". *Journal of Rural Studies* 28: 199–207.

6 Geographical indications – a double-edged tool for food democracy

The cases of the Norwegian geographical indication evolution and the protection of stockfish from Lofoten as cultural adaptation work

Atle Wehn Hegnes and Virginie Amilien

Introduction

In July 2002, the regulation and labeling scheme for protected designation of origin (PDO), protected geographical indication (PGI) and traditional specialities guaranteed (TSG) was set up in Norway modeled on the corresponding sui generis system for geographical indications (GIs) in the European Union. The Norwegian government's aim was to prevent copying of product names, to increase the variety of foodstuffs available, to protect and revitalize local production and know-how and to add value to Norwegian food production. The scheme was also intended to provide sufficient information and transparency to consumers. As of February 2019, twenty-nine Norwegian and three Italian products have been granted PDO, PGI or TSG in Norway.

The implementation of a system for GIs in Norway was demanding, requiring administrators, producers, consultants and others to make a significant and all-round effort to adapt the scheme to the Norwegian food culture and, conversely, to adapt Norwegian food culture to the scheme. The aim of this chapter is to probe this mutual cultural adaptation work (CAW) (Hegnes 2013), its power relations and its consequences by addressing whether GI is a universal tool for local socioeconomic development and democratization of agri-food and fish products.

By probing the power dynamics and adaptation practices necessary to implement and use the global system of GIs in the specific Norwegian national and food-cultural context, our chapter relates to the body of literature that seeks to understand and develop the concept of food democracy (FD). FD is a *sensitizing concept* (Blumer 1954, 7) that has been underdeveloped since it was coined by Tim Lang in the late 1990s (Lang 1999a, 1999b). Various authors have discussed its content, focusing on different aspects such as food security (e.g., Sonnino, Moragues Faus and Maggio 2014), power relationship between consumers and market or producers (Dubuisson-Quellier and Lamine 2008;

Tovey 2009), the global food system (Lang 1999a, 1999b) and citizenship and human rights (Hassanien 2008; Lamine, Darolt and Brandenburg 2013). Booth and Coveney claim that "The essence food democracy is about the redistribution of power within the food system" (2015, 16). Hence, the conceptual and empirical approaches of FD are strongly interdependent, and changes in social practices are necessary for FD to function as a pragmatic method for transforming the agri-food system (Hassanein 2003). However, to understand the complexity, facets and dynamics of FD is demanding. Renting, Schermer and Rossi emphasize that

> What is needed especially are conceptual approaches that address more clearly the renewed role in such dynamics of citizens, consumers, producers and civil society, the distinctive nature and characteristics of social and economic relations embodied in newly emerging food networks, and their potential to generate genuine food system transformations.
>
> (2012, 292)

In this regard, we argue that CAW represents a conceptual approach that can contribute to understanding parts of the dynamics that Renting, Schermer and Rossi (2012) address.

A general description and understanding of the implementation and use of the GI system in Norway makes up the first and overarching case in this chapter, whereas *Tørrfisk fra Lofoten* (Stockfish from Lofoten (SfL)) is used as a specific case of a GI product. SfL is selected as unit of analysis mainly because it is also registered as a third-country GI product in the European Union. Including the Norway/EU dimension makes it possible to consider not only the local and national levels but also the multilevel dimension and complexity of GI systems as part of the analysis – making the power within, and the consequences of, the adaptation work even more complex and intriguing.

In the next section, we go into detail on our sources of data and methodological approach. We also outline our conceptual framework for describing and understanding the power dimension in GI systems. In the third section, we describe and analyze the Norwegian historical and contemporary context and its impact on the implementation and use of the GI. In the fourth section, we analyze the construction of quality and the evolution of protection of SfL in Norway and the European Union. In the last section, we conclude by discussing the overall adaptation work, its power dynamic and its intended and unintended consequences on GI as a tool for FD.

Methodological approach and analytical framework

The methodological approach of this chapter is qualitative, and the analysis is based on diverse forms of empirical material, its origin in two different studies: Hegnes (2013) and the EU-funded project Strength2food.[1] The data were collected and produced between 2008 and 2018, and they consist of material such as document studies of laws, regulations, policy documents and interviews with people responsible for working out product regulations in producer organizations. Interviews have also been conducted with key informants representing public bodies that administer the regulation. The interviews were all semi-structured.

As mentioned, the goal of FD as a pragmatic political tool is to change the dominant food system (Hassanein 2003, 77). Renting et al. claim in their argument that civic food networks (CFNs) and food citizenship are core elements in building FD and that "CFNs present highly diverse forms of civic engagement in food networks and in their transformative potential within agri-food systems and, more generally, food practices and culture" (2012, 299). Thus, the dynamics of change is important to address in order to meet the ambition of change and key dimensions in FD (Hassanein 2008, 289–291).

Several approaches can be used to interpret whether GI is a suitable tool for local socioeconomic development and the democratization of agri-food. Here, we propose to understand the potential of GI for FD through an analysis focusing on the power dimension in food-cultural adaptations and adaptive practices. According to Hegnes, the food-cultural perspective allows us to understand food culture(s) as continually changing processes (2013, 19). How actors in different contexts adapt the schemes for GIs to the culture and the culture to GIs can be understood and defined as different *adaptation practices*. The practices vary depending on what is to be adapted, which means that it can involve a variety of practices. Hegnes focuses on three ideal-typical adaptive practices and describes the dynamics between them (2013, 96–98):

Translations are the work exerted to create changes and adaptations of *meaning* in Norwegian food culture relating to the labeling scheme and product regulations. Translation occurs when new meanings arise or are being changed. Through the adaptation work the meanings of places, products and brands, menus and so on are constructed.

Reorganizations are changes in *social relations and groups* or the creation of new groups. Reorganization can occur both at the individual, group and nation levels. In addition, reorganization occurs in real groups, such as the formation of producer associations and groups on paper,[2] as they are constructed by researchers, market analysts, politicians and others. Both forms of reorganization are important, and there is interaction between them. Real groups give rise to groups on paper, and vice versa.

Transformations are innovations or changes of *material conditions*. Transformation is related to the material aspect of labeling schemes and mainly the products. In addition to the products, means of production and a more general understanding of nature can also be taken as a basis for transformation.

The actors' overall adaptation work can be understood as the sum of the interdependent practices that occur in the interaction between human translations

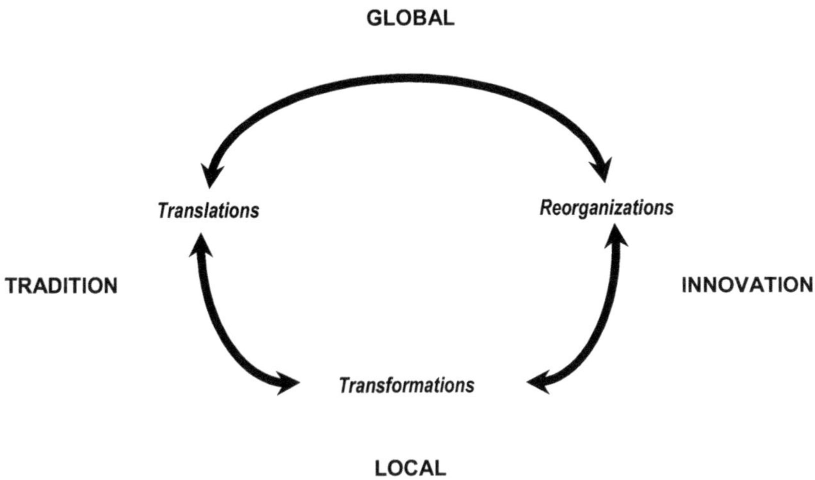

Figure 6.1 Dimensions and dynamics in CAW

Source: Hegnes 2013, 25.

of meaning, reorganization of social relations and transformation of things (materiality) in the tension between the global and the local and between tradition and innovation, as shown in Figure 6.1.

As already mentioned, GIs are intended to protect producers, in terms of both knowledge and production methods/heritage. De jure, producers should have power to decide on their participation and prescribe their products. De facto, the systematization of GIs at the EU level means that progressively lower stakeholders in the hierarchy are potentially subordinate when they adapt to the system. Applying CAW as a heuristic approach will allow us to "identify, understand and describe how the adaptations take place, who carries them out, what characterizes them, and what consequences and implications they have" (Hegnes 2013, xiv).

Introducing and adapting geographical indication to the Norwegian context

Whereas the Norwegian GI system was developed and adapted according to the common European system, the EU system was inspired by existing national systems, such as the French AOC (*Appellation d'Origine Contrôlée*) and the Italian DOC (*Denominazione d'Origine Controllata*). The French AOC scheme is closely related to the concept of terroir, which dates back to the 13th century (Bérard and Marchenay 2008; Barham 2003). In short, a terroir product is characterized by a specific geographical origin, developed over a long period of interaction with the local traditions, local environment and savoir faire. Such factors are considered in applications for an AOC/PDO/PGI label. Although the importance of terroir has changed and continues to evolve (Amilien 2011), it is beneficial that French manufacturers, the French government and consumers have a more or less common understanding of the term's content. In Norway, there is no equivalent term that emphasizes the dynamic trinity of product, people and a specific place, but this does not mean that *terroir products* did not exist in Norway (Hegnes 2012).

Pre–geographical indication introduction

Norwegian food traditions had a modest impact during the 19th century nation-building process, compared to other parts of Norwegian folk culture (Notaker 2000, 2001; Amilien and Stø 2000, 10). After World War II, farmers' cooperatives in milk, meat and poultry effectively controlled these markets. During 1970–1980, the grocery distribution in Norway was largely an independent wholesaler industry. The distribution channel for groceries was basically open to all suppliers, and the individual merchant decided the specific range of groceries for his shop. During the 1980s, the concentration of the retail and the vertical integration between retail and wholesale increased. In combination with a soft discount domination of the market, this led to a restrictive assortment policy. Until this point, the quality focus was mainly on the *national* and not the regional or local level (Amilien and Hegnes 2004).

Due to international negotiations and agreements in the 1990s, Norwegian food products faced increased competition from foreign imports. To counter the competition, Norwegian authorities and other key stakeholders started constructing what some described as "mental border protection." Simply put, this strategy aimed to convince Norwegian consumers to choose Norwegian products, also termed *gastronationalism* (DeSoucey 2010). From this point, Norwegian governmental authorities have promoted a range of different food qualities and the sale of high-quality food through different labeling schemes. This new regime triggered a novel way of thinking about and looking at food that emerged in Norway toward the end of the 1990s and beginning of the 2000s.

The Norwegian turn to new qualities coincides with a growing focus on food labeling in Europe over the past two to three decades (Hegnes 2012). This European "quality turn" indicates a turn made by consumers, retailers and producers away from standardized products toward alternative qualities (Goodman 2003). In the Norwegian Agricultural Authority's strategy work at this time, it was a clear goal to base the "mental border protection" on Norwegian competitive advantage. Among such advantages were explicitly mentioned local food traditions and old Norwegian food traditions/food culture and the like (Norwegian Ministry of Agriculture 1992).

Whereas some "quality turns" can be understood as bottom-up processes, the Norwegian one was characterized by a top-down initiative. This difference implies that what is referred to as the "quality turn" may, in different contexts, cultures and countries, consist of entirely different ways of turning to and constructing alternative qualities (Hegnes 2012; Halkier, James and Stræte 2017). Introducing and adapting the GI scheme and products is an important example of the initiative and mechanisms in the Norwegian turn to new qualities.[3]

Post-geographical indication introduction

A hierarchical division of the administrative responsibility of the Norwegian GI system was organized during the implementation. The Ministry of Agriculture and Food is ascribed the overall responsibility for the system, while the Norwegian Food Safety Authority and the independent Norwegian Food Branding Foundation are attributed to other "lower" areas of responsibility.

Since their introduction in 2002 until today, the Norwegian scheme and its regulation for GIs have been revised a number of times. The revisions are interesting and are concrete examples of how actors initiate and perform food-cultural adaptation work by adapting to several conditions. While the introduction of the regulation is often referred to as having been done by following the pattern from the European Union, it became apparent already in the first revision in 2003 that it is also translated according to patterns in Norwegian food culture.

The first revision resulted in the inclusion of fish and fishery products in the scheme. There were various views on including fish, but the potential of branding SfL in the export market, registered as a third-country product under the European scheme, was an important and strong motivation for the change. What initially seems to be a "narrow" translation to Norwegian conditions changes and expands when we consider that stockfish is also related to the southern countries, which account for a large proportion of the consumption of stockfish from Norway. In this context, the complexity of the food culture and the food-cultural adaptation work becomes evident. The Norwegian regulation is actually adapted to the Norwegian *and* Italian food cultures at the same time – more specifically to producer groups in the Lofoten archipelago above the arctic circle and Italian consumers in the Northern part of Italy.

More generally, the development of the Norwegian GI scheme strongly underlines that GI is not only a technical structure but also a political one. Within this structure, there is a power struggle between different actors, such as when decisions made by the producer groups are closely related to policy decisions, which are *translated* into important internal discussions in producer groups (Fonte and Grando 2006).

The most important adaptation practices in the introductory phase of GIs in Norway are *translations* and *reorganizations*. The adaptation of meaning through translations takes place in the implementation of the European GI system in the Norwegian context. Reorganization was also important in achieving a management model for the new qualities and the new scheme, adapted to Norwegian food culture and the Norwegian context.

A specific vocabulary or language makes it possible to communicate the food-cultural understanding that forms the basis for the scheme in the European

Union. This understanding involves, inter alia, that the relationship between tradition and the food's place of origin is considered meaningful and important in the product's quality. In Norway, there is lack of both the vocabulary and food-cultural know-how in line with the food-cultural preconditions that form the basis for the scheme in the European Union. The work of establishing and *translating* a corresponding vocabulary and understanding in Norwegian food culture has therefore been particularly important during the introduction, administration and use of the scheme. However the terroir concept does not just travel between food cultures through different kinds of translations (Barham 2003; Cappeliez 2017). The concept must also be understood as a food-cultural category that brings about other adaptive practices. We will now elaborate on this in the marine terroir story of SfL.

Adapting stockfish from Lofoten to Norwegian and European geographical indications

StL is made of winter cod ("skrei") fished around the Lofoten islands, carefully cleaned and prepared before it's wind dried during the ideal temperatures of the winter months. SfL is one of the oldest Norwegian export products and definitively an old and famous part of Norwegian cultural heritage. Already in the early Middle Ages, it was exchanged and sent to other European countries by the Vikings. In 2007, it was granted a PGI in Norway, and in 2014, it was the first Norwegian product to receive a European PGI.

In 1998, the Norwegian Seafood Council (the export office aiming at marketing fish products)[4] ordered an analysis of the Italian dry fish market (see Mangseth and Teigland 1998), which revealed a strong interest in and recognition of SfL in Italy. The idea of developing a GI was later discussed among the producers of SfL. However, they did not come to an agreement – on the one hand, because their own networks and markets were well established and, on the other hand, because of existing tensions between some of the local producers.[5]

After SfL was protected in Norway in 2007, with the acknowledgment of "Beskyttelse av Tørrfisk fra Lofoten som geografisk betegnelse,"[6] several actors cooperated to submit an application for SfL as a third country in the European Union.[7] First, a request was sent to Innovation Norway,[8] aiming at getting financial support for rewriting the specifications and adapting them to both English language and the specific requirements of the European Commission. The revised English version of the application was sent to the European Union on 8 November 2012, submitted for consultation on 11 December 2013 and eventually approved on 22 April 2014.

Since it was first proposed in 1998, nearly twenty years passed before an agreement on, and the fulfillment of, the technical obligations of the GI were realized. Despite information and support from the Norwegian Food Branding Foundation, this registration period dragged on due to difficulties in understanding the law, gathering and organizing the consortium and collecting all the necessary documentation. The dynamic processes behind strategies of GI boost awareness among actors engaged in food production and preparation and encourage other actors (government representatives, export council, NGOs, etc.) to take an active role in the promotion of the product, including its culinary specificities and qualities. To understand the mechanisms behind this dynamic, we use CAW as a heuristic model. First, we focus on the history of export to describe the contextual frame in which the quality before PDO/PGI was considered, and eventually, we observe the impact of PGI as a quality scheme.

The adaptive evolution of stockfish export

Historically, the export of stockfish has long been one of the main resources for Norwegian society. In *Egil's Saga*[9] we can read about trade with England around 875 as "Thorolf had this ship made ready. . . . And freighted it with dried fish and hides, and ermine and gray furs too in abundance, and other peltry such as he had gotten from the fell; it was a most valuable cargo" (*Egil's Saga* 1240, chapter 17).

Stockfish has always been produced in the northern part of Norway. The Lofoten archipelago has played a central role because of its special climate and natural conditions. As of 2018, three-quarters of the total volume of stockfish comes from the Lofoten archipelago. In the 1300s, several orders of the king emphasize the importance of the production of dried fish in Norway, which brought a large amount of money to the kingdom. However, between the end of the 1200s and 1715, there was no direct trade between foreigners and the Lofoten Islands, because foreign vessels were not allowed to go up north to collect fish (as King Håkon Håkonsson had conferred on Bergen the monopoly of trade to the north). Foreigners had to stop in Bergen, and the city became a fundamental link not only in stockfish trade but also in the construction of the stockfish quality grade. While the quality of the stockfish from Finnmark (in the most northern part of Norway) was defined by its weight, the SfL (south of Finnmark) was differentiated by the German Hanseatic tradesmen by its length, taste and leanness to obtain different prices.

In the Middle Ages, stockfish was exchanged against clothes, wine or wheat from the southern part of Europe and sent to England, the Netherlands and the Vatican or to France for centuries. But both Italy and Nigeria also played a major role in the more recent development of trade for SfL over the last century, with fluctuations from period to period. The main differences between Italy and Nigeria are related to the trade agreements and the quality dimension. First, the two markets do not build on the same premises, as the stockfish trade with Africa is based on the exchange of stockfish against palm oil, including prices strongly regulated by states (Konow 1945, 22–24), while the pillar of the Italian market is tradesmen, who include such agents as producers, exporters and importers. Second, West-African countries imported, until recently, types or parts of stockfish that the Italian market was not demanding, such as heads (as West-African eaters cooked stockfish in stews mixed with red pepper). In Italy, stockfish was a quite regular product, bought and consumed by common households as a sort of low-price food. However, it had to be of high quality since it was in competition with local products such as eggs or meat. Stockfish is found in numerous parts of Italy, often following several recipes and ways of preparation and consequently using different types and quality of dried fish. Especially in the northern part of Italy, where thin fish is appreciated, the quality of SfL has a higher status than standard stockfish from Norway (Konow 1945, 17), while the consumers in southern Italy use a cheaper and thicker type of fish (Martinussen, Richardsen and Ulve 2000, 200). Quality was an imperative part of the Northern Italian market for stockfish, which was long led by a few well-structured and efficient exporters in Bergen (Egils Saga 1240, 68), with the consequence that, since the end of the 1980s, Northern Italy has been the main importer of SfL. But Lofoten is nowadays experiencing a "quality turn" and a new approach to stockfish from the Nigerian market interested in higher-quality products.

This brief historical overview can be understood as *translations, reorganizations and transformations*. First, we have seen that previously each country had its own translation of the term *tørrfisk*, sometimes based on the original meaning of the word (i.e., dried fish) and on phonetics, such as "stockfish" or the Italian "stoccafisso." Nevertheless, we also find other concepts, such as the *baccala*, from Piero Qverini, an Italian artisan from Venezia who ended up for several months in Lofoten in the winter of 1432 after a shipwreck. The story of Qverini is used at a basic reference in the GI's application because of the memory of his Norwegian travel preserved in the Vatican Apostolic Library (Querini 1533). *Baccala* is still used in the Venice area, often mixed with *baccalao*, and back in Norway, underlining a dynamic influence of translation. Second, our overview shows the necessity of a strong reorganization of social networks and previous agreements, such as the state regulation for the Nigeria market. The relation between traders and importers for the Italian market is also reconsidered through the lens of the consortium. Third, new ways of transporting the dried fish, by road and not by sea for example, did necessitate transformation and innovation impacting the fish consumption. To elaborate further on the FD-GI-CAW nexus, we focus specifically on the evolution of the quality dimension.

Pre-protected geographical indication quality

In terms of technical concepts, the quality of PGI SfL is defined in market and statistical material as "round fish from Lofoten."[10] However this broad definition actually includes several types of quality. Although the two main quality categories of stockfish are defined as "first (prima)" and "second (secunda)" (as shown in Table 6.1), Martinussen, Richardsen and Ulve (2000, 199) describe altogether twelve quality categories already used by the Hanseatic commercial system and based on the market place, length, weight, quantity of fish per

Table 6.1 Older quality categories for *Stockfish from Lofoten*

Name – either describing the fish or referring to a market place	*Quality criteria Size – length or quantity of fish per kg*
Variants within prime quality – stockfish from Lofoten	
Ragno	More than 60 cm
Westre Magro	50/60 cm 60/80 cm
Westre Demi Magro	50/60 cm 60/80 cm
Grand Premier	60/80 cm
Westre Courant	75–80 fish per 50 kg
Westre Piccolo	100–120 fish per 50 kg
Westre Ancona	75–80 fish per 50 kg
Hollender	58–63 fish per 50 kg
Bremen	50–55 fish per 50 kg
Lub	40–45 fish per 50 kg
Variants within second quality – stockfish from Lofoten	
Italia Grande	50–60 fish per 50 kg
Italia Grande Magro	60–65 fish per 50 kg
Italia Medio	75–80 fish per 50 kg
Italia Medio Magro	75–80 fish per 50 kg
Italia Piccolo	100–120 fish per 50 kg

Source: Authors' elaboration on data available in Martinussen, Richardsen and Ulve 2000.

kilogram, as well as the sound of the dried fish when it was knocked (a feature which is not in the table).

Importers and consumers from Northern Italy have a deep knowledge of the products from Lofoten. "Ragno," a thin round fish, has always been highly praised, as leanness is a mark of delicacy in the northern part of Italy. Until the 1900s, all discarding (quality grading) of SfL took place in the export ports, mainly in Bergen. Today the grading history is described in the PGI regulation to highlight the increasingly objective quality assessment:

> The knowledge of grading is vital to the quality of the end product. Grading requires in-depth knowledge and is done by so-called "discarders". Historically speaking, the quality assessment is a skill which has been handed down through generations in families and/or companies. Formerly, graders were trained in the company with which they were associated, while lately theoretical grader training has been established for the industry. . . . This knowledge is essential in order to obtain good quality of "Tørrfisk fra Lofoten".[11]

In this perspective, the hierarchy of qualities was actually reduced already before SfL obtained a PGI by simplifying the old Hanseatic system of *prima* and *secunda* types of fish, based on size and weight. Furthermore, before the PGI, the name of "the Lofoten archipelago" functioned as "a mark of quality" itself. Organizations of trade networks, between Norwegian and Italian families, or agents and producers, based on trust and knowledge, were both endorsing and reinforcing this mark of quality.

Post-protected geographical indication quality

Translations turned out to be central also after the PGI, as social practices, and particularly practices concerning the market and distribution abroad sometimes develop a different understanding of the stockfish concept, and consequently, a second translation of the original term can be a major issue in defining quality. In Italy, for example, where the term *stoccafisso* appears to be the literal translation, the secretary of the PGI *tørrfisk fra Lofoten*'s consortium noticed that most Italian dealers use the name which designates the best quality of dried fish, *ragno*, as the standard name of dried fish. Thus, the term *ragno* became synonymous with dried fish when it originally meant a high-end product. This translation problem creates an imbalance in the Italian market, where the different qualities are no longer communicated and valued by consumers, whereas the prices are originally quite different. The secretary of the PGI consortium therefore concludes that everyone must stop speaking "their own language" and that the different Norwegian and Italian actors must agree on a common language.

> Italians use a term that does not correspond to Norwegian content of the term. . . . They talk about Ragno, but it really means just dried fish. It does not matter about quality. So Stockfish producers should have a discussion and see if we need to spend so much time and energy to have so many categories. . . . maybe we should go down to 3 categories. We also wonder if we would have our own category called "dry fish from Lofoten PGI."[12]

While the linguistic aspect is quite easy, as everyone adapted and appropriated the Norwegian concept of *tørrfisk*, with reference either to the quality of the product (dried fish) or to its form (stock) in different languages and pronunciation, the translation of knowledge into juridical terms was far more challenging. The authors of the product regulation specifications had to translate local knowledge and culture as well as know-how of the fisheries and process of drying and qualifying cod into a technical language. "Atlantic cod captured around Lofoten and Vesterålen from January through April" had to be defined in concrete geographical terms and the "territories" to be delimitated on a map, as local ways of describing the fishing place, were not detailed enough in the Norwegian product regulation.

In the Norwegian regulation for PGI for SfL[13] we can read on the second section of the chapter about the geographical area: "Lofoten consists of the municipalities Flakstad, Moskenes, Røst, Vestvågøy, Værøy and Vågan." In the European regulation for the PGI, we can read in the second section about geographical area a detailed description of the place (e.g., "between Ø 010°00' to Ø 016°08' and N 67°00' to N69°30' in the following ICES-location"), with concrete geographical references illustrated by a map.[14]

When it comes to natural conditions, the Norwegian regulation for the PGI contains three or four sentences about the perfect climate for drying fish and the unicity of the place. Otherwise, the European product regulation is made more scientific and technic and turned to be almost one page,[15] beginning with "Temperature, precipitation, wind, sun and snow are all factors vital to the drying process. Because of the Gulf Stream, winters in Lofoten are mild with an average temperature of between -0.8 to +2.2 °C from January to April, and dry spring with a total precipitation of 132 to 108 mm."

The most challenging translation is obviously the grading part, as experts recognize the quality of the stockfish from the sound the fish makes when it is knocked. This tacit and incorporated know-how is traditionally learned by doing, from generation to generation. It is demanding to have it translated into technical concepts and words. The European specification on SfL took up the challenge, and we can read that standardization and determination of normative standards for manufacturing and grading have come in recent times. Recently, there has also been established discarder training for the industry, which consists of theoretical education as well as practical training in the companies.[16]

The quality dimension was also adapted and restricted to "a golden colour on the skin and a size between 40–90 cm." Because of the PGI specifications, the consortium decided to use the size and thickness of the fish and to adapt to the Norwegian standards:[17]

> Tørrfisk fra Lofoten of the Prima category shall be virtually flawless with a natural shape and open belly. . . . Tørrfisk fra Lofoten of the Second category can have small, but not significant flaws.
>
> Tørrfisk fra Lofoten is to be sorted into various quality groups according to the Norwegian Industry standard for classification of stockfish, NBS 30–01. Important criteria for further grading into quality groups are the size/weight and thickness of the fish.[18]

We also note a material *transformation* as a result of the work on agreeing to the product regulation. Stockfish was usually stored on wooden pallets and sold "just like this," but will now be "packed in jute or cardboard." The PGI product obviously has to be differentiated from the non-PGI, but the material transformation can vary: as some Italian consumers call it, "the one packed in plastic with black lettering."

Moreover, the GI for a product which developed through several centuries of export tradition necessarily demanded a *reorganization of social relationships*.[19] The cooperation within the PGI consortium changed traditions and routines for several producers who were used to having their own networks and carrying out their own business practices. All members had to collaborate on the PGI, at least through the consortium but also with other actors. A few producers of SfL are still not in the PGI consortium, because they do not want to join it. However, in the near future, the consortium will increase from eighteen to twenty members.[20] The PGI consortium covers about 80 percent of all stockfish producers in Lofoten. There are mainly fish processors, as sixteen of them buy fish and often produce stockfish as a part of their business, but also two companies buy stockfish only from other producers and trade with national and foreign actors.

The PGI consortium, together with other stakeholders, markets actors and regional public organizations, aims for a better use and promotion of the SfL PGI. The PGI consortium also plans to improve the use of the PGI label in the Italian market, discussing alternative models for cooperation with actors on the Italian market.[21] This will certainly change the landscape of the traditional relationships between Norwegian and Italian actors. Moreover, they also plan *transformations*, especially innovation, by thinking about increasing the processing phase and proposing consumers get "ready to eat" products instead of full dry fish that needs to be irrigated (eight to twenty days) before it can be eaten. Nevertheless, the PGI label and the PGI as a marketing tool are still slightly visible, and it seems that most producers still use their traditional network.

To sum up this short description of the evolution overview of stockfish quality, we see that the several-centuries-old SfL also had to go through at least three types of adaptation. The works of *translations* are to be seen at the local level, because the product itself got a new meaning by its public acknowledgment; at the national level for similar reasons but essentially because of the value of national cultural heritage; and eventually at the international level, where translations directly affect the product definition itself. The works of *reorganizations*, which concern changes in social relations and groups, are especially interesting at the local level, where producers have to cooperate and agree, while they actually have been competing with each other for centuries. The reorganization of the "foreign link of the distribution chain", including Italian sellers, fishmongers and Italian consumers, also changed after the new definition of quality imposed by the European quality schemes. The relationships between researchers, food authorities, communal and national stakeholders and producers must also be considered as a dynamic of change, especially because the whole process of applying for the GI began after a research study ordered by the Innovation Norway office. Getting a PGI also demands several innovations of *material conditions* or works of *transformations*, like the new packaging (including a differentiation of packaging between local consumers, national consumers, tourists in Norway and foreign consumers in foreign countries), the development of new ready to eat products and a reconsideration of the traditional criteria for fish quality.

Conclusion

As this book shows more broadly, adapting the GI system to foreign or historically unfamiliar institutional and cultural conditions is increasingly commonplace as GI expands to new territories. Over the last two decades, GIs have been strongly promoted outside of their "traditional" Western frame, not only to support rural development but also to be more visible on the global quality food market. As Calboli puts it,

> In particular, countries in Asia, Africa, and South America have adopted, to date, national policies on GIs and taken part in the discussion of the WTO Doha Development Agenda. In the past decade, provisions related to GI protection have also become an important component of bilateral and plurilateral international trade agreements (FTAs) between countries from all continents.
>
> (2017, 4)

In this chapter, we have addressed whether GI is a universal tool for local socioeconomic development and the democratization of agri-food and fish. Based on our general description and analysis of the Norwegian GI system and the product-specific case of SfL, our conclusion is that GI is not such a universal tool. Once it's been adopted, it needs to be adapted. The tailoring to specific food-cultural contexts and the necessary adaptation work further leads to intended and unintended consequences.

We have shown that the Norwegian GI evolution and the protection of SfL can be understood as a chain of adaptations and adaptive practices. The adaptations are necessary to unite the dynamic ordering of modern global WTO regulations and national food systems with the food-cultural status of traditional local products. Our approach, to analyze the potential of GI as a tool for FD, has thus been in line Hassanein's claim that "Food democracy is about citizens – in the broad, denizen sense of the word – determining agri-food policies and practices locally, regionally, nationally, and globally" (Hassanein 2008, 289).

We have shown that the consequences of the adaptation work are, to varying degrees, in line with the ambitions of the scheme. Our study shows that new vocabulary and new food-cultural knowledge makes it possible to communicate and understand links between food, people and place, which have previously not been emphasized, such as Norwegian food culture. It appears, however, that the adaptation work also promotes innovation and sometimes alienates producers from their own products over the course of the application process, which conflicts with the aim of preserving knowledge. The adaptation work also includes adjusting the product name. Moreover the demarcation of geographical origin and the issue of borders (Hegnes, forthcoming) are also questions of adaptation. We noticed how the coastal boundaries were "translated" and delimited from the national to the European application. This is consistent, to varying degrees, with the intention of giving sufficient information to consumers and preserving important knowledge about Norwegian food culture. Another consequence of the adaptation work is that it plays a part in products becoming more similar, which again results in fewer product types or qualities. This also conflicts with the scheme's aim of contributing to an increased variety of foodstuffs.

As a whole, the scheme is often described as being modeled according to the European system. However, understanding it as a consequence of the adaptation work emphasizes the dynamic of the scheme, which appears to be a tailor-made Norwegian model with a European profile. Balance is then a key issue: If the scheme is excessively adapted to the Norwegian context, it may risk losing its credibility and validity in the European Union. If the Norwegian system is excessively adapted to the European scheme, it may be perceived as cumbersome and meaningless for Norwegian producers, retailers and consumers.

In matter of food democracy (FD), one of the most important impacts of the quality schemes of the SfL is undoubtedly at the communication level. Both the process linked to the application for a PGI and the acceptance create an interesting and quite transparent arena for knowledge exchange. The fact that only a few actors take part in the debate is nevertheless problematic as it makes the process exclusive, although this is obviously not due to an undemocratic system but instead to a lack of interest (especially from the consumer side). A similar mechanism is also apparent at the system level. The influence of the European scheme system on the Norwegian one is generally less about the expression of power and more about its passive structuring role, or model of power (Bråten 1973). However, regardless of whether the power mechanisms are actively expressed or passive structuring, their effect is important the potential of GI as a tool for local socioeconomic development and the democratization of agri-food and fish.

The food-cultural adaptation work of GIs gives voice to and empowers local actors and subordinated groups, but it can at the same time be seen as an instrument that hampers democratic forms of development. The implementation and establishment of GIs in Norway have been parts of a food-cultural evolution over the last twenty years, when focus on the combination of food, people and places has played a central role. In this last perspective of enlightenment, the process is a step toward better common knowledge, which is an implicit pillar of democracy. GI may thus be a tool for local socioeconomic development and democratization of agri-food and fishery if the schemes are carefully adapted to the national food culture and if the food cultures are carefully adapted to the scheme. However, in sum, so far, the Norwegian GI scheme and the case of SfL show that GI is a double-edged tool for local socioeconomic development and FD.

Notes

1 The Strength2food project received funding from the European Union's Horizon 2020 research and innovation program under grant agreement No. 678024.
2 Groups on paper can be understood to be in line with Bourdieu and Hacking's understanding of, respectively, "classes on paper" (Bourdieu 1985, 725) and "making up people" (Hacking 2007).
3 This section is partly based on Hegnes and Gustavsen (2017).
4 Quotation from their webpage: "The Norwegian Seafood Council (NSC) aims to increase the value of Norwegian seafood resources. We do this through market insights,

market development, market risk management and reputational risk management in select markets around the world." https://en.seafood.no/about-norwegian-seafood-council/about-us/ (Accessed: 19 September 2018)

5 Personal interviews with members of the consortium in 2005 (Fondation Franco-Norvégienne project) and in 2018 within the Strength2food project.

6 The original law from 11 December 2007, FOR-2007-12-11-1814, was revised as FOR-2012-06-29-694.

7 Interview with Nina Wærnes Hegdahl, lawyer responsible for GIs at Matmerk (4 October 2017).

8 An agency who "is the Norwegian Government's most important instrument for innovation and development of Norwegian enterprises and industry." www.innovasjonnorge.no/en/start-page/

9 The *Egil's Saga* manuscript is a family Icelandic Saga from 1240 CE, which documents the life of the farmer and viking Egil Skallagrimsson and his family.

10 For example, in http://seafood.no/markedsinnsikt/apne-rapporter/manedsstatistikk/

11 Regulation (EU) No. 1151/2012 of the European Parliament and of the Council on quality schemes for agricultural products and foodstuffs. http://eur-lex.europa.eu/LexUriServ/LexUriServ.do?uri=OJ:C:2013:361:0010:0012:EN:PDF

12 Interview with Anne Karine Statle, 29 November 2017. min. 53.15

13 Forskrift om Tørrfisk fra Lofoten FOR-2007–12–11–1814

14 Regulation (EU) No. 1151/2012 of the European Parliament and of the Council on quality schemes for agricultural products and foodstuffs. http://eur-lex.europa.eu/LexUriServ/LexUriServ.do?uri=OJ:C:2013:361:0010:0012:EN:PDF

15 Ibid.

16 Ibid.

17 Information from the PGI consortium underlines that they are working for a third category, which could be "other," but one that only the PGI product could use. (Author's interviews, 29 November 2017).

18 Ibid.

19 Interview with Nina Wærnes Hegdahl, lawyer responsible for GIs at Matmerk (4 October 2017)

20 Personal interviews with Rune Stokvold, chairman of the Tørrfisk fra Lofoten consortium, 2 January 2018, and Anne Karin Statle, secretary of Tørrfisk fra Lofoten and project leader of "Lofoten Mat", a network of local food producers, 29 November 2017.

21 Personal interviews with Rune Stokvold, chairman of the Tørrfisk fra Lofoten consortium, 2 January 2018.

Bibliography

Amilien, Virginie. 2011. "From Territory to Terroir? – The Cultural Dynamics of Local and Localized Food Products in Norway." *Sosiologisk Årbok* 16 (3/4): 85–106.

Amilien, Virginie, and Atle Wehn Hegnes. 2004. "The Cultural Smell of Fermented Fish – About the Development of a Local Product in Norway." *Food, Agriculture & Environment* 2 (1): 141–47.

Amilien, Virginie, and Eivind Stø. 2000. *Om matkultur . . . Et felles SIFO-arbeid på oppdrag fra Statens Landbruksbank* (Arbeidsnotat nr. 19–2000). Lysaker: SIFO.

Barham, Elizabeth. 2003. "Translating Terroir: The Global Challenge of French AOC labeling." *Journal of Rural Studies* 19 (1): 127–38.

Bérard, Laurence, and Philippe Marchenay. 2008. *From Localized Products to Geographical Indications. Awareness and Action.* Bourg-en-Bresse: CNRS Ressources des terroirs.

Blumer, Herbert. 1954. "What Is Wrong with Social Theory?" *American Sociological Review* 19 (1): 3–10.

Booth, Sue, and John Coveney. 2015. *Food Democracy* (SpringerBriefs in Public Health). Singapore: Springer Singapore.

Bourdieu, Pierre. 1985. "The Social Space and the Genesis of Groups." *Theory and Society* 14 (6): 723–44.

Bråten, Stein. 1973. "Model Monopoly and Communication: Systems Theoretical Notes on Democratization." *Acta Sociologica* 16 (2): 98–107.

Calboli, Irene. 2017. "Geographical Indications Between Trade, Development, Culture, and Marketing: Framing a Fair(er) System of Protection in the Global Economy?" In *Geographical Indications at the Crossroads of Trade, Development, and Culture: Focus on Asia-Pacific*, edited by Irene Calboli and Ng-Loy Wee Loon, 3–35. Cambridge: Cambridge University Press.

Cappeliez, Sarah. 2017. "How Well Does Terroir Travel? Illuminating Cultural Translation Using a Comparative Wine Case Study." *Poetics* 65: 24–36.

DeSoucey, Michaela. 2010. "Gastronationalism: Food Traditions and Authenticity Politics in the European Union." *American Sociological Review* 75 (3): 432–55.

Dubuisson-Quellier, Sophie, and Claire Lamine. 2008. "Consumer Involvement in Fair Trade and Local Food System: Delegation and Empowerment Regimes." *GeoJournal* 73 (1): 55–65.

Egils Saga. ca. 1240. Trad 1893 Translation into English by W. C. Green from the Original Icelandic *'Egils Saga Skallagrímssonar'*. Online. http://sagadb.org/egils_saga.en.

Fonte, Maria, and Stefano Grando. 2006. *A Local Habitation and a Name – Local Food and Knowledge Dynamics in Sustainable Rural Development*. 2nd Version. WP6 CORASON – Local Food Production Comparative Report, August 2.

Goodman, David. 2003. "The Quality "Turn" and Alternative Food Practices: Reflections and Agenda." *Journal of Rural Studies* 19 (1): 1–7.

Hacking, Ian. 2007. "Kinds of People: Moving Targets." *Proceedings of the British Academy* 151: 285–318.

Halkier, Henrik, Laura James, and Egil Petter Stræte. 2017. "Quality Turns in Nordic Food: A Comparative Analysis of Specialty Food in Denmark, Norway and Sweden." *European Planning Studies* 25 (7): 1111–28.

Hassanein, Neva. 2008. "Locating Food Democracy: Theoretical and Practical Ingredients." *Journal of Hunger & Environmental Nutrition* 3 (2–3): 286–308.

———. 2003. "Practicing Food Democracy: A Pragmatic Politics of Transformation." *Journal of Rural Studies* 19 (1): 77–86.

Hegnes, Atle Wehn. forthcoming. "The Map and the Terroir – Adapting Geographical Boundaries for Geographical Indications in Norway." *British Food Journal.*

———. 2013. "Kulturelt tilpasningsarbeid. Innføring, forvaltning og bruk av merkeordningen Beskyttede betegnelser i Norge (Cultural Adaptation Work – Implementation, Administration and the Use of Geographical Indications in Norway)." Doctoral thesis, Department of Sociology and Human Geography, Faculty of Social Sciences, University of Oslo.

———. 2012. "Introducing and Practicing PDO and PGI in Norway – Turning to Protected Quality Through Translations of Meaning and Transformations of Materiality." *Anthropology of Food* 57, online document retrieved at: http://aof.revues.org/7210 on 4 May 2019.

Hegnes, A. W., and Gustavsen, G. W. 2017. "The Class and Culture of Norwegian Culinary Straw Men: A Response to Flemmen, Hjellbrekke and Jarness' 'Class, Culture and Culinary Tastes: Cultural Distinctions and Social Class Divisions in Contemporary Norway.'" *Sociology* 53 (3): 600–08.

Konow, Carl. 1945. *Tørrfiskhandel – en faglig utredning om Tørrfiskhandel fra 1914 til 1945.* Bergen: AS John Griegs boktrykkeri.

Lamine, Claire, Moacir Darolt, and Alfio Brandenburg. 2013. "The Civic and Social Dimensions of Food Production and Distribution in Alternative Food Networks in France and Southern Brazil." *International Journal of Sociology of Agriculture & Food* 19 (3): 383–401.

Lang, Tim. 1999a. "The Complexities of Globalization: The UK as a Case Study of Tensions Within the Food System and the Challenge to Food Policy." *Agriculture and Human Values* 16 (2): 169–85.

———. 1999b. "Food Policy for the 21st Century: Can It Be Both Radical and Reasonable?" In *For Hunger Proof Cities: Sustainable Urban Food Systems*, edited by Mustafa Koc, Rod Macrae, Luc J. A. Mougeot and Jennifer Welsh, 216–24. Ottawa: International Development Research Center.

Mangseth, Lise and Kristin Reichel Teigland. 1998. *Analyse av det italienske tørrfiskmarkedet: foredragsnotater Publisert ved Ballstad i Lofoten: Eksportutvalget for fisk.* https://urn.nb.no/URN:NBN:no-nb_digibok_2014070308117.

Martinussen, Terje, Roger Richardsen, and Stein Ulve. 2000. "Norsk eksport av tørrfisk til Italia." *Økonomisk Firskeriforskning* 10 (2): 199–211.

Norwegian Ministry of Agriculture. 1992. *Competitive Strategies for Norwegian Food* (The clean food strategy). Oslo: Ministry of Agriculture.

Notaker, Henry. 2001. "Nasjonal matkultur." *Nytt Norsk Tidsskrift* 18: 205–6.

———. 2000. "Har vi en nasjonal matkultur?" *Nytt Norsk Tidsskrift* 17: 345–64.

Querini, Pietro. 1553–59. "*Navigationi et viaggi*-in Memory of His Norwegian Travel: The Story Is Preserved in the Vatican Apostolic Library Cod. Vat. Lat. 3256." Here Referring to the Norwegian Translation in Wold H. A 1991 *I paradisets første krets*, Cappelens Forlag.

Renting, Henk, Markus Schermer, and Adanella Rossi. 2012. "Building Food Democracy: Exploring Civic Food Networks and Newly Emerging Forms of Food Citizenship." *International Journal of Sociology of Agriculture and Food* 19 (3): 289–307.

Sonnino, Roberta, Ana Moragues Faus, and Albino Maggio. 2014. "Sustainable Food Security: An Emerging Research and Policy Agenda." *International Journal of Sociology of Agriculture and Food* 21 (1): 173–88.

Tovey, Hillary. 2009. "Local Food as a Contested Concept: Networks, Knowledges and Power in Food-Based Strategies for Rural Development." *International Journal of Sociology of Agriculture and Food* 16 (2): 21–35.

7 The decline of the French label of origin wine

Romain Blancaneaux

Introduction

Under neoliberal globalization, geographical indication (GI) is often regarded as a tool for revitalizing agricultural communities through the reappropriation of local spaces, cultures and practices and through granting higher incomes to their users. However contemporary, this idea is historically rooted in the emergence of the concept of certifications for place-based products. As a product of the first half of the 20th century, the establishment of the denomination of origin (DO – in this chapter used as a form of GI) wines in France stabilized the market and restored confidence after decades of uncertainty and fraud (Garcia-Parpet 2009, 9–21; Stanziani 2003a, 2003b). Differentiating them from common wines and placing them under the interventionist supervision of the state because of chronic overproduction, the institutionalization of DO wines created a model for excellence in standards and promotion while allowing producers to control their own regulation. It was based on a conception of quality that stressed the importance of "a term untranslatable in any other language, terroir: a proof of a social and historical specificity, evoking equally qualities of the soil, the climate, and of traditional practices [that lend the market] a stable social structure" (Garcia-Parpet 2009, 14). For wine producers who agreed to comply with collectively set regulations for traditional practices and quality, both in their vineyards and their wineries, the DO appellations seemed to offer levels of sales and incomes commensurate with their extra effort and know-how (Unwin 1996, 315–316). For most of the 20th century, these kind of GI wines[1] abided by the exclusive orientation of this concept, such that by 1970, GI wines were produced in relatively modest amounts – covering only 20.1 percent of the national vineyard area and constituting 15.4 percent of production (Humbert 2011, 571, 580) – and sold in small quantities (3 liters per capita per year) (Arnaud 1991, 6). The scarcity and singularity of GI wines arose out of a narrow historical structure of the wine market, the classification of 1855 (*classement de 1855*), which circumscribed GI wine along the predilections of experts in the wine sector with shared social standing and distinctively historical production practices (Garcia-Parpet 2009, 15).

In the early 1970s, the sale of GI wines was carried out through traditional small-scale distribution rather than large-scale logistics (Bartoli and Boulet 1989, 554). Too expensive to ensure sufficient stock and profitability for large-scale retailers, GI wines, or the "'grand crus' wines," had been nearly absent from their shelves (Bartoli and Boulet 1989, 556). For vine growers, however, access to a GI label was seen as a consecration, a guarantee of economic and symbolic success that would only increase with the age and the reputation of the vineyard (Garcia-Parpet 2009, 12, 20) as long as traditional standards were upheld. These conditions allowed consumers to develop confidence in the DO concept and encouraged producers to reinforce the use of the label (Fourcade 2012, 528). By shaping the behavior of producers and consumers alike, the GI label created a commercial institution that was firmly in place by the time the European Economic Community (EEC) adopted a regulatory framework for wine. Following common regulations and procedures shared by other sectors, wine was controlled by a common market organization (CMO); the wine CMO was established based on preexisting national institutional structures and regulations designed to monitor all aspects of wine production. By 1970, and through intergovernmental discussions at the EEC level, French negotiators and experts were able to win regional recognition for their national legislation and regulation. In fact, the EEC modeled European categories of wines after the French concept of quality wine produced in specific regions (QWPSR) (Smith, de Maillard and Costa 2007, 80).

About forty years later, the GI model of excellence inspired by French GI wines has surprisingly been destabilized, including in many areas where it originated. In contrast with the 1970s, when purpose and motivation seemed self-evident, a growing malaise has emerged in recent years toward labeling that has affected GI wine producers and consumers alike (Garcia-Parpet 2009, 11). After periods of historically low prices and their wider availability at discount and large-scale distribution outlets, sales of GI wine have reached a level of mass consumption not discernible from that of table wines (ONIVINS 1994, 140). Indeed, shares of table wine in total production had fallen from 85 percent to 48 percent between 1969 and 1992, while GI wines had become dominant (ONIVINS 1998, 61). Perhaps because of this "success," a growing segment of certified GI wine producers began to feel that the use of this label no longer provided adequate remuneration. "Some French producers do not feel rewarded, neither by the work methods nor by the farming choices prescribed by the historically established quality standards such as the '*classement de Bordeaux* of 1855,' and geographical indications" (Garcia-Parpet 2009, 11). An increasing number of winegrowers are shifting to table wine production and abandoning the GI label that, ironically, was designed to serve as a better alternative to this type of wine. For the past ten years, some table wines have generated such good returns to former GI wine producers that many who produce a local-type wine (uncertified but following various appellations: department, zone or varietal) can command higher prices than comparable GI wines (de Cantenac 2005, 65). As stated by the French Interprofessional Office for Wines

(*Office National Interprofessionnel des Vins*, ONIVINS), the public institution in charge of the wine sector in France: "There is thus no more immediate alignment for the consumer between spontaneous perception of respective qualities of the various categories of wines, and their price. It is estimated today that a third of the GI production is in direct competition with table and local wines in the domestic market" (ONIVINS 2003, 16).

At the European level, the GI-based classification between wine categories has proven itself to be obsolete and overwhelming for consumers (4,200 GIs for wines in Europe, 466 in France alone), and poorly aligned with the World Trade Organization (WTO) Trade-Related Aspects of Intellectual Property Rights (TRIPs) agreement that the European Union signed in 1994 (Allaire et al. 2005, 3). Accordingly, in order to meet and protect "consumer-oriented measures," the European Union proposed a "simplification of wine classification" through a new scheme bearing "clearer, simpler and more transparent rules as well as WTO-TRIPS compatibility," which was to be implemented through CMO reform (European Commission 2006). In 2007, the French negotiators and experts accepted this scheme, abandoning the system that their peers had helped establish decades earlier. This situation raises questions about the integrity of the GI label and the excellence associated with GI wine production. What happens when GIs grow perpetually? What if a substantial share of the products is labeled in ways that become increasingly incomprehensible to the public and international market? What happens when the exclusiveness of GIs is replaced by inclusivity? Because of its long-term evolution, the case of French GI wines provides a background for questioning GI as a tool for managing commoditization and democratically controlling the impacts of neoliberal globalization on agri-food. This comes into relief in the case of French wine, where GI was framed well in a domestic, national context but grew increasingly complex, contradictory and ineffective as the consumer base expanded to the rest of Europe and the world, where the framework provided by the WTO has been prevailing.

This chapter offers a compelling counterargument that even in the bastions of GI policy – the iconic case of French GI wine – GI does not readily represent a safeguard against the negative effects of neoliberal globalization. Furthermore, the evolution of French GI wines speaks to the paradoxical outcome that expansion of GI products is not a linearly positive or virtuous development but rather an accumulation of stakeholders whose strategic behavior must be managed by evolving institutions. By examining the tensions between the national and international scales of action, this chapter tackles a wider problem.

While one may contend that the raw proliferation of GI wines is the decisive factor in the deterioration of quality associated with the DO label, it would be erroneous not to consider the effects of the scope of the regulation itself. As developed by microhistorians (Revel 1989, 1996), the effects-of-scales method permits us to see how rationales deployed at a smaller spatial level (France) do not produce the same results at a larger spatial level (Europe). Following this framework, the research in this chapter is based on the consultation of existing documents, such as professional journals and official texts, and twenty semi-structured interviews conducted between 2011 and 2014 with current and former officials of the European Union, the French Ministry of Agriculture, the French Fraud Service and Control and the National Institute for the Denomination of Origin (INAO).

The chapter contends that the emergence of GI initially fortified wine producers against liberal measures and state intervention. In the 1970s, in the nascent European Community, the French concept of DO provided the basis for shaping the common regulatory framework thus protecting wine producers' institutional independence and intellectual property. In the following two decades, the expansion of, and demand for, the production of French GI wines went hand in hand with the international validation of the French concept. However, beginning with the 1990s, a dilution of DO principles and gradual subordination to the French state, and the European Union can be seen as the price paid for the "success" of the GI. Simultaneously, a change in the production and reception of French expertise on wine-related issues has cast into doubt the potential for public GI institutions to objectively and fairly serve as arbiters of agri-food heritage.

The development of "place of origin" as a market norm against economic liberalism and state intervention

In the first half of the 20th century, GI wines rose to prominence as a response to decades of rampant economic liberalization, market instability, weak profit margins and exposure to risk that public interventions had been unable to mitigate. In France, this situation engendered a specific notion of GI that was later adopted throughout the European Union.

A reaction against unprecedented economic and symbolic upheaval

In the 18th century, wine quality was associated with the accumulated reputation of international merchants whose actions conferred on their wines important symbolic and economic capital. The growth of this class of merchant was linked to the handling of structural and political conditions, such as shipping costs and export tariffs, that conditioned wine trade at the time. This situation prompted segments of the agrarian bourgeoisie to invest in winemaking techniques that eventually permitted the production of quality wines at desired prices. The net result was that this quality production achieved higher enough prices to offset the shipping costs and tariffs that affected the wine sector. In France, the Bordeaux-area wines benefited most from this move (Butel 1963; Pijassou 1970). Their reputation was ratified through the *classement de 1855* that, employing local names (or *crûs*) for the first time, differentiated superior quality wines from lower-grade wines. This local-name-based differentiation bestowed upon these wines a quality status that implicitly pitted them against other types of wines.

By the middle of the 19th century, however, the liberalization of trade and production practices associated with processes initiated by laissez-faire liberalism gradually upended the status quo. Additionally, the diffusion of railroads and the lowering of international trade tariffs contributed to the rationalization of the wine market inside and outside of France. These structural changes prompted the intensification of trade in ordinary wine production as local producers found ordinary wines economically more attractive than *crûs* wines. In that period, powdery mildew and phylloxera epidemics further engendered a rise in prices that helped even ordinary wines achieve high price points (Cadier 1972, 21). As these epidemics caused wine shortages, unconventional winemaking practices developed to exploit high prices, such as employing dried fruits or chemicals to inflate production volumes (Stanziani 2003a, 2003b).

The development of GI arose, in this nexus, as a reaction to the liberalization of wine production and trade. Although the phylloxera crisis ended by the turn of the century, alternative winemaking practices persisted, reducing outlets for genuine grape wine producers, let alone high-quality *crûs* wines. Moreover, as some wine merchants tried to unfairly benefit from the fame of classified *crûs* (Stanziani 2003b), conflict emerged with wine producers. They organized in unions and called for the state to implement legal measures to protect "natural wines" (*vins naturels*), made of fresh grapes, from "false wines" (*faux vins*) (Stanziani 2003b, 164). Included in these demands were measures to prevent the swapping of wine and/or grapes from lesser-known areas to renowned regions. In 1905, the government passed the Law on Frauds and created the Fraud Control Service and the Designation of Regional Origin (*Appellations d'Origine régionale*) to protect quality wines. Even then, the limited success of the implemented measures became apparent in the continued conflict between merchants and local producer unions, the inability of local appellate courts to address escalating conflicts and the multiplication of unwarranted appellations. Progressively, this resulted in a regionalization of viticulture control.

As wine professionals sought to manage their markets and maintain independence from ineffective state interventions, a coalition of stakeholders promoted the creation of a new category of wines, known as the Designation of Origin (*Appellations d'Origine Contrôlée* or AOC). Although the name was not dramatically different from predecessor appellations, the governance was. This new system, created in 1935, was regulated by a centralized, professional body known as the *Comité National des Appellations d'Origine des Vins et Eaux-de-Vie* or CNAO, which was renamed in 1947 as the *Institut National des Appellations d'Origine Contrôlée* (INAO), and was endowed with the ability to certify wine producers and their products and to regulate markets (Garcia-Parpet 2004, 74).

International endorsement of specificities

Created as an opposition to the liberalization of markets and haphazard state intervention, the establishment of the DO regulatory system rested on the fundamental contribution of different categories of experts, principally agronomists, lawyers and geographers (Roger 2010, 1096–1098). Among them, lawyers had long positioned themselves at the center of the process of "construction of quality," starting with the 1851 Act on a More Effective Repression of Certain Frauds in the Sale of Goods, with oversight at different levels of the juridical system (Stanziani 2003b, 173). Repeated falsifications of quality led lawyers to push for measures against "false wines," which led to the 1905 concept of Denomination of Regional Origin and the establishment of the Fraud Control Service. In the wake of the government's inability to effectively define the areas eligible for GI wine production, lawyers reemerged to play a central role in appellate courts' efforts to define local "fair and traditional practices" in 1919 (Stanziani 2003b). By supplementing their legal credentials by teaming up with agronomists and geographers, these experts reinforced their position and legitimacy within the field and provided more differentiated instruments for protecting regional specificities. Furthermore, CNAO and later INAO were allowed to maintain autonomy from the state and wine producers. They could draft statutory orders and recruit their own agents (agronomists and lawyers alike) without government approval and interference, even as far as to place fraud agents under its brigade of control of DO (*brigade de contrôle des appellations d'origine*).

As the EEC was established in the late 1950s, differences among the GI systems became not only evident but also a source of contestation (Barthe 1988). Compared to the French, the German system was not as centralized, and its quality adjudication scheme was based exclusively on the degree of alcohol content present in wine. The Italian structure, for its part, was by far the most liberal among those of the wine-producing countries and characterized by the presence of only a few appellations. In the meantime and facing limited alternatives, French experts were able to strongly position their conception of DO wine, not only at the International Organization for Vine and Wine (OIV) – an institution concerned with technical and scientific aspects of winemaking – but also at the "Lisbon agreements" of 1958 (Jacquet 2005). Given the progressive success in achieving international recognition, the French conception of DO wine was eventually adopted by the EEC as well. Simultaneously, the production and sales of French GI wines grew significantly.

Productive and institutional expansion: the strength of the French conception of denomination of origin

The newly formed EEC emerged as a favorable ground for the legitimation of French legal experts. Simultaneously, the French concept of DO wines remained popular despite the decline of wine consumption.

A sprawling network of French legal experts

As the EEC was formed, its Wine Division had to formulate a CMO for the sector. To ensure the EEC's operational capability and legitimate its expertise and authority, civil servants were recruited among administrators and professionals familiar with the wine sector regulations (Scheinmann and Feld 1972, 123). These recruits tended to maintain close ties with their home bureaucracies, and these bureaucrats brought with them "baggage of preconceived ideas, outlooks and prejudices, many of them of a specific national nature" (Cockfield 1994, 109; Scheinman 1966, 761). This form of recruitment allowed individual member states to maintain various levels of influence that encouraged sectorial and national cultures to take root in the EEC (Balint, Bauer and Knill 2008, 683). One result was the development of a shared understanding between French representatives and European civil servants that ultimately allowed the French DO concept to prevail in the Wine Division. This result, however, was not preordained.

The first statutory texts for the organization of the European wine sector brought together national delegations eager to defend their views and domestic systems. The important status held by the French Fraud Service, INAO and the Ministry of Agriculture allowed their representatives not only membership in the French delegation but also significant influence to sway the European model toward the nascent French model of DO wine. The first step, however, was to ensure that the domestic French system would emerge from the negotiations unscathed. Ultimately, the Treaty of Rome (1958) stated that a supranational market organization could replace a national organization only if it guaranteed a framework equivalent to existing regulations (Treaty establishing the European Economic Community 1957, articles 39, 42, 43).[2] The Treaty basically left the French GI wine organization untouched under the newly created European category of quality wines produced in specified regions (QWPSR). The tacit approval of the French conception of DO wines at the EEC reinforced France's prominent position at the supranational level.

Propelled by the liberalization of trade within the EEC and the use of French appellations by other countries, the French conception for DO continued to gain prominence in Europe. One impact that emerged in the 1970s with the lowering of tariffs within the EEC, and the geographical growth of the market thanks to European economic unification, was the increased availability of foreign counterfeits. This motivated INAO's Legal Service and the EEC alike to prioritize the protection of GI wines. One key measure was to enforce that the bottling of GI wines occur in producing regions, which became mandatory in 1974.[3] And in a more general sense, the Ministry of Agriculture's Fraud Control Service and INAO strengthened their leadership in legal expertise by contributing to lawmaking and acting internationally. This was particularly the case for INAO concerning Champagne.[4] In 1985, to thwart foreign attempts to alter the use and understanding of GI, INAO and the Wine Bureau of the Fraud Control Service created the International Wine Law Association (AIDV) within the OIV. Eventually, both INAO and the Ministry's experts became highly respected as they occupied important positions within the EEC and Permanent Representative working groups (Scheinman 1966, 759–760).

Additionally, relations with the EEC's Wine Division were maintained by "staffing it with already trained and well-placed [French] civil servants," which created a sort of "parallel administration" known formally as "Seconded National Experts" (Scheinmann and Feld 1972, 122; Trondal 2003). This situation further cemented the French administrative culture within the EEC. In the midst of this, and on until the late 1980s, the number and volume of French GI wines continued to grow (Abélès, Bellier and Mcdonald 1993, 40; Smith 2001).

The growth wave of national geographical indication wines

By the time the EEC legislation on wine endorsed the French conception of DO, French GI wines were enjoying a period of strong growth as a share of all household beverages domestically. However, between the 1960s and the late 1980s, wine consumption declined as a share of household beverage expenditures from 68.4 percent in 1987 to 50.8 percent in 1960. This drop, however, was felt primarily by ordinary wine producers, while GI wine consumption continued to grow throughout this period. Measured as a proportion of household wine expenditures, GI wine consumption grew from 16.4 percent in 1960 to 26.6 percent in 1970, 30.1 percent in 1980 and 36.0 percent in 1987 (Bartoli and Boulet 1989, 565, 632). The growth in consumption extended also to various intermediate quality categories, which also contributed to the relative decline in the consumption of ordinary wine.

Beginning with the 1970s, the expansion of fine wine consumption was paralleled by a shift in their mode of commercialization that penalized traditional distributors and favored large-scale retailers (Daumas 2006, 57). As stated by a sectorial observer, "DO wines initiated a remarkable commercial breakthrough in the large-scale retail sector" (Réjalot 2007, 45). Up to that time, large commercial retailers carried primarily ordinary table wines. As demand for fine wines increased, they quickly assigned greater portions of their shelf space to GI wines because they were showing better margins than table wines (Bartoli and Boulet 1989, 556; Réjalot 2007, 53). By the early 1980s, large-scale retailers distributed almost half of the production of fine "authentic" wines (47.3 percent) and continued to expand their availability as supply also increased. In this context, declining consumption and relatively high handling costs signed the economic fate of table wines. The greater public exposure of GI wines motivated winemakers to ramp up investments that would ensure higher and more consistent production volumes. Accordingly, production rose from 4 million hectoliters in 1960 to 8 million hectoliters in 1980, with an increase in land devoted to GI wine of approximately 60,000 ha between 1970 and 1985 (Bartoli and Boulet 1989, 16, 365). This increase in production subsequently generated a price reduction that further stimulated demand (Bartoli and Boulet 1989, 651). The wider accessibility and mainstreaming of GI wines, however, began to chip away at its original appeal, namely exclusivity.

Shifting trends under a new EU authority

As the European Union was formed, the success of GI wines provided pretext for the French Ministry of Agriculture to take direct control of the wine GI. Subsequent intervention by the French government was paralleled by stricter EU control of GI and the subordination of many forms of geographical designation to TRIPS.

Subordination to the ministry of agriculture

The success of GI wines prompted the French Ministry of Agriculture to employ it as a model to protect and promote other quality food products, a move that would lead to negotiations with the European Union. In the early 1990s, the French government (acting through the Ministry of Agriculture) pushed to extend the jurisdiction of INAO to include additional agricultural products.[5] INAO was reorganized to include three committees with a director providing oversight in the following categories: Wine and brandy; cheese, milk products, and other agri-foods; and meat. This structural change altered the role previously assigned to legal experts in defining GI products. While intending to simply extend the wine conception of DO to other products, the specificity of wine proved awkward for other agri-food sectors. Accordingly, the unwieldy GI criteria were progressively abandoned and replaced by more universal norms aligned with INAO's Legal Service, which, by that point, was largely staffed with lawyers from the Ministry of Agriculture. They brought with them different understandings of quality already existing in other sectors – such as Label Rouge for meat and cheese products – and pushed to implement different legal definitions of the GI label to achieve their sought-after "multi-product" evolution.

By the mid 1990s, emboldened by an unprecedented market downturn, the Ministry of Agriculture acted to curtail INAO's traditional authority in GI wine. At the time, increases in cultivated area and production had saturated the market. Production of DO wine alone had increased from 15 percent of the total wine production in 1969 to 52 percent by 1992, even as the national vineyard area almost doubled from 1970 (243,439 ha) to 1993 (427,089 ha) (ONIVINS 1993, 23). Paralleling this production boom were growing price disparities and quality issues that cast a shadow on the DO market (L'Union girondine des vins de Bordeaux 1993, 3). GI wines without a strong reputation were increasingly competing with table wines that had previously been seen as categorically inferior (Montaigne 1994, 5). In some cases, DO wines lost favor in whole regions. In Languedoc-Roussillon in 1995, for example, the price per hectare for table wine vineyards exceeded the majority of GI hectarage (ONIVINS 1995, 72). As faith in the self-regulating capacity of the DO wine market evaporated, the Ministry took over. By this time, consumption of GI wine had reached mass consumption levels. Productivism reached such a point that even the INAO director, Alain Berger, denounced it (Maleysson 1995). Confronted with this set of problems, INAO experts were discredited, and their institution underwent reform.

The net result of this process was that the Ministry of Agriculture gradually turned INAO into a multi-product labeling agency following EU guidelines. In this context, the implementation of new GI registration procedures involved the development of internal relations and competencies in accordance with the European Union's directives for trade-related issues (Charlier 2007, 73). By 1999, the Ministry of Agriculture extended the duties of INAO to include all quality specifications covered by European regulations, while a newly created fourth Committee for agri-food was staffed with officials of the Ministry. By 2005, there were 593 registered French GI products, 466 of which were wines (Allaire et al. 2005, 3).

A takeover by the European Union of foodstuff quality signs

In the first wave of GI institutionalization in the nascent European Union, the French were able to play a leading role in validating their national model, but eventually the direction of influence turned against the French. By initially responding to the European Union's intention to place all national foodstuff quality signs under a single jurisdiction, the French Ministry of Agriculture helped INAO to establish a leading role in the shaping of European regulations on GI products. This allowed the French to deflect the first potential dilution of the DO concept by engineering an exclusion for wine, which was considered to have more specificity. The eventual compromise led to the two-category European system still in operation today among EU member states. The first category is the protected designation of origin (PDO) based on the French concept of terroir, which was placed under INAO supervision in France. The second, the protected geographical indication (PGI), which was more based on the typicity of the production of agri-food, was placed under the jurisdiction of the Ministry of Agriculture.

Beginning in the 1980s, the European Union worked to progressively affirm its authority over trading issues as it considered them falling under its jurisdiction. For one, the European Union attempted to shore up its limited supranational authority to regulate PGIs and PDOs of all member states. In the run-up to the Maastricht Treaty, the European Union hoped to "clearly establish or reinstate its authority as the sole negotiator" on international commercial issues. These moves, however, failed to gain acceptance from member nations (Billiet 2006, 907). Ultimately, the creation of the WTO addressed the issue in that it reinforced the European Union's scope of international trade negotiations by creating a centralized dispute settlement mechanism and pushed for consistent worldwide intellectual protections under TRIPS. In February 1996, the European Union launched a new Market Access Strategy, which relied on the use of the WTO's dispute settlement system to protect intellectual property rights as defined in TRIPS. GIs were included in this process (Hudec 1993; Billiet 2006, 901). This change coincided with a discursive reorientation of the European Union's wine division toward more horizontal, conformist conceptions of GI.

By the late 1990s, an internal process of reorganization had altered the European Union's interest in intellectual property rights and finally led to a weakening of the dominance of the French conception of DO wine. In part, this arose out of the new meritocratic "open career system" adopted in European institutions, whereby European civil servants were increasingly required to be internationally focused in their skills and outlook (Balint, Bauer and Knill 2008, 684; Stevens 2002). This new recruitment procedure brought about a fundamental shift in individuals' work orientation, with interest growing in extending the GI system into new territory rather than qualitative regulatory control at the national level. The Wine Division's officials began to engage in direct, bilateral agreements with third countries to expand the GI system globally. However, the multiplication of these agreements and the sheer number of GIs – 4,800 registered, among which 4,200 were for wines and six hundred for other products (Allaire et al. 2005, 3) – contradicted claims of qualitative exclusivity and actually served to discredit the European GI policy ex officio in WTO panels (Petiteville 2006, 165). Controversy peaked in 1999, when the United States challenged the European Union in the WTO, contending that the patchwork of national registration procedures under the European GI system was inconsistent with its obligations of harmonization under TRIPS. The WTO dispute settlement panel ruled in 2005 in favor of the United States. Following the suit, the European Commissioner for Agriculture, Mariann Fischer Boel, agreed to amend the regulations by consolidating quality foodstuff products in a "harmonized" framework compatible with the TRIPS.

Conclusion

Historically, wine has been the first food product for which regional origin and differences increasingly gained relevance among producers and consumers, to the point of giving birth to a quality standard, the appellation of origin (*appellation d'origine contrôlée*). The concept, formerly conceived for wine, was the first kind of GI instituted nationally that established an "'exclusive' attribution of the characteristics of a product to its place of origin" (Parasecoli 2017, 31), which is now understood as the stricter category of GIs. The genesis of the French GI for wine echoes the overall goal of this volume, namely to track the potential of GIs to contest neoliberal globalization by promoting local socioeconomic development and democratization of agri-food.

Indeed, born in reaction to economic and commercial structural changes between the 19th and 20th centuries, which hastened liberalism and threatened the incomes and reputation of famous wine, GI provided the market with stability, and producers with better incomes overall. Local producer groups could take control of their heritage agricultural systems to raise incomes while increasing the verifiable diversity and quality of the wine selection. This model in France was so effective initially that other countries followed the system by introducing similar regulations, indicating that the strict constellation of terroir not only provided marketing appeal but helped to preserve local heritage while maintaining quality. However, the success of the French wine GI, as traced in its contemporary evolution in this chapter, challenges the assumption that the institutionalization of GI is a linearly positive progression.

First, this chapter questions a common assumption of GIs as a tool for achieving qualitative differentiation of food products. As Fabio Parasecoli notes, differentiation spurs along the basic process of market segmentation, wherein consumers find a wider range of products whose certified specificities match their expectation of quality and price (2017, 46). As far as French wines are concerned, the creation of GI came as an answer for producers in renowned winegrowing regions to counteract the perception of free riding by producers of common wine, who unduly took advantage of – and thus jeopardized – the symbolic and economic capital of famous producers by producing wines of lesser quality. By delimiting geographic areas of production and standards of quality, and thereby restricting the number of agents and the volume of wine on the market, GI regulations aimed at avoiding heterogeneous wine specificities that oversaturate the market and bring prices down. GI wines differentiate themselves by simultaneously legitimizing their terroir and achieving high enough prices that consumption and cultural reproduction is limited exclusively among the upper classes. The stricter the standards of quality, the higher the financial and social barriers posed to producers and consumers to access the GI market. For a time, this worked well for consumers, who could trust the framework of regional producers operating under the objective professional oversight of CNAO, established already in 1935 (later renamed INAO). It also worked for producers, who could more autonomously govern production standards in regional groupings and thereby escape from ineffective national level interventions. Finally, the GI also worked for nascent European authorities, who, by 1970, in the establishment of the Wine CMO, largely confirmed the French terroir-based model of wine quality differentiation. Gradually, however, the differentiation process of GI wines progressed to such a degree that the GI ceased to represent higher quality and instead simply represented a wide diversity of qualities.

This evolution in France suggests a fundamental contradiction of GI concerning the extent of exclusivity. On the one hand, small producers rely on the exclusiveness of their products to justify higher prices and reputation. On the other hand, GIs are a form of "collective good" that should not be denied to qualifying producers, which makes them theoretically inclusive and thus unable to maintain exclusiveness over the long term. What happens, then, when GIs actually become more inclusive? Here, the stunning evolution of French GI wines over the past forty years offers a cautionary tale about how, despite being borne of a will to forge local autonomy and differentiate themselves, the landscape of GI wine invariably became so overly differentiated that it began to intersect with table wine. By 2005, some GI wines competed in price and reputation against wine types which had once been valued the least. Against all expectations, GI wine gradually went from structurally marginal to dominant in global production. This was echoed within France, as the continuous extension of vineyard area, and the intensification of its production progressively led GI wines, once the exception, to become the rule. Although this can be seen as potentially emancipating for everyday consumers, who are able to access cheaper, albeit uneven quality, GI wines, this has been to the detriment of a growing margin of producers. Such an evolution offers an example of how the value made by a GI initiated a cat-and-mouse game that incentivized more production and subsequently its mainstreaming, as well as poor self-discipline in managing its "exclusivity."

Because this occurred to wine in France, it puts into question the potential for exponential expansion of GI in the world. GI is now commonly considered a tool for economic growth, and its diffusion to the intellectual property regimes of much of the world during the past decade has been buoyant. The quick implementation in some poor agricultural countries has fostered hope for securing local know-how, curbing the circulation of imitation goods and ensuring better revenues against domestic and international competitors, in a hostile neoliberal regime. What is to happen, ultimately, if the number and volume of GIs grows exponentially worldwide? What happens when the labeling of "exclusivity" becomes increasingly incomprehensible to the public and the international market? Here, the case of French GI wines offers a cautionary tale and the beginning of an answer. The higher domestic and international commercial value of French GI wine has not only encouraged winegrowers to grow preexisting GIs more intensively on larger areas but created a diffusion of new GI wines, which peaked at 466 GIs in 2005. While inflation caused oversupply, heterogeneity in qualities and reputation also served to lower prices and the general standing of GI wine. This heady expansion of GI wine has continued even as the European Union began working to harmonize its intellectual property regime under the TRIPS, with the effect that the current producer-unfavorable GI mechanism is effectively France's marquee model of GI institutionalization.

Indeed, the evolution of French GI wines contradicts the belief according to which expertise can be sustainably afforded by local professionals and legitimated in a disciplined and consistent way by national professionals. The national level was once the source for producing wine-related expertise, which rested on a historical monopoly of INAO and its legal experts working in concert with local producer associations. This system was confidently put forward as a model for European GI regulation schemes even as it was destabilizing domestically in France. This destabilization hastened dramatically from the 1990s, when a multiplicity of events converged to progressively empower the European Union on matters of external trade prerogatives and foodstuff quality signs. Changes in the European Union's interests in international negotiations coincided with an evolution of the French wine institutions that diminished the role of the INAO experts who had previously disciplined the liberalization of the wine sector. Indeed, the setting up of European regulations on quality agri-food products, coupled with the French wine sector's economic hardship, pushed the French Ministry of Agriculture and INAO to cede their policy-making to the framework established by the WTO.

The long lifespan of the French wine GI and its various institutional and legal shifts bring up new questions about the long-term capacity of GIs to serve as a tool for local socioeconomic development and the democratization of agri-food under neoliberal globalization.

Notes

1 This book is about GIs. This is the most general and widely known categorical name for these types of certifications. Because of the international orientation of this book, we choose later on to privilege the term "GI" instead of other formulations such as DO, albeit anachronical.
2 Regulation (EEC) No 817/70 of the Council of 28 April 1970 laying down special provisions relating to quality wines produced in specified regions, § 15.
3 Council Regulation (EEC) No. 2133/74 of 8 August 1974 laying down general rules for the description and presentation of wines and grape musts, Official Journal of the European Communities, No. L 240, July 08, 1976, p. 1–20; Commission Regulation (EEC) No. 1608/76 of June 4, 1976 laying down detailed rules for the description and presentation of wines and grape musts, Official Journal of the European Communities, No. L 183, 8 July 1976, p. 1.
4 See Paris Court of Appeal, 1st room A., 15 December 1993.
5 Act No. 90–558 of July 2, 1990 on Designations of Origin for Agricultural or Food Products, Raw or Processed, OJ of 6 July 1990, pp. 7912–7914.

Bibliography

Abélès, Marc, Irène Bellier, and Maryon Mcdonald. 1993. *An Anthropological Approach to the European Commission*. Brussels: Report for the European Commission (unpublished).

Act No. 90–558 of July 2, 1990 on Designations of Origin for Agricultural or Food Products, Raw or Processed, OJ of July 6, 1990, pp. 7912–14.

Allaire, Gilles, Bertil Sylvander, Giovanni Beletto, Andrea Marescotti, Dominique Barjolle, Erik Thévenod-Mottet, and Angela Tregear. 2005. "Les dispositifs français et européens de protection de la qualité et de l'origine dans le contexte de l'OMC: justifications générales et contextes nationaux." Symposium international Territoires et enjeux du développement régional, Lyon, mars 9–11.

Arnaud, Charles. 1991. "Le vin et l'organisation commune de marché entre Paris et Bruxelles: un dialogue quelquefois difficile." *Économie rurale* 204: 3–10.

Balint, Tim, Michael Bauer, and Knill Christoph. 2008. "Bureaucratic Change in the European Administrative Space: The Case of the European Commission." *West European Politics* 31 (4): 677–700.

Barthe, René. 1988. *L'Europe du vin: 25 ans d'organisation communautaire du secteur viti-vinicole: 1962–1967*. Paris: Ed. Cujas.

Bartoli, Pierre, and Daniel Boulet. 1989. *Dynamique et régulation de la sphère agro-alimentaire. L'exemple viticole*. Montpellier-Paris: Économie et Sociologie rurales, INRA.

Billiet, Stijn. 2006. "From GATT to the WTO: The Internal Struggle for External Competences in the EU." *Journal of Common Market Studies* 44 (5): 899–919.

Butel, Paul. 1963. "Grands propriétaires et production de vins du Médoc au dix-huitième siècle." *Revue historique de Bordeaux*, 129–41.

Cadier, Gabrielle. 1972. "L'exportation des vins de France sous le second Empire." *Bulletin du centre d'histoire économique et sociale de la région lyonnaise*, 10–26.
Charlier, Christophe. 2007. "La protection européenne des indications géographiques face au principe du traitement national de l'OMC." *Économie rurale* 299: 70–83.
Cockfield, Lord. 1994. *The European Union. Creating the Single Market.* Chichester: Wiley Chancery Law.
Commission Regulation (EEC) No. 1608/76 of June 4, 1976 Laying Down Detailed Rules for the Description and Presentation of Wines and Grape Musts, Official Journal of the European Communities, No. L 183, July 08, 1976, pp. 1–6.
Council Regulation (EEC) No. 2133/74 of August 8, 1974 Laying Down General Rules for the Description and Presentation of Wines and Grape Musts, Official Journal of the European Communities, No. L 240, September 3, 1974, pp. 1–20.
Daumas, Jean-Claude. 2006. "Consommation de masse et grande distribution. Une révolutionpermanente (1957–2005)." *Vingtième Siècle. Revue d'histoire* 91 (3): 57–76.
De Cantenac, Pierre. 2005. "Vins de pays. Le Languedoc s'envole." *La RVF* 489.
European Commission. 2006. *Towards a Sustainable European Wine Sector.* Brussels.
———. 1996. *The Global Challenge of International Trade: A Market Access Strategy for the European Union.* Brussels.
Fourcade, Marion. 2012. "The Vile and the Noble: On the Relationship Between Natural and Social Classifications in the French Wine World." *The Sociological Quarterly* 53 (4): 524–45.
Garcia-Parpet, Marie-France. 2004. "Le marché de l'excellence: le classement des grands crus à l'épreuve de la mondialisation." *Genèses* 56 (3): 72–96.
Garcia-Parpet, Marie-France. 2009. *Le marché de l'excellence. Les grands crus à l'épreuve de la mondialisation.* Translated by the author. Paris: Seuil.
Hudec, Robert E. 1993. *Enforcing International Trade Law: The Evolution of the Modern GATT Legal System.* Salem, New Hampshire: Butterworth Legal Publishers.
Humbert Florian. 2011. *L'INAO, de ses origines à la fin des années 1960. Genèse et évolutions du système des vins d'AOC.* Translated by the author. Dijon. PhD Thesis, Université de Bourgogne.
Jacquet, Olivier. 2005. "De la Bourgogne à l'International: construction et promotion des normes d'appellation d'origine ou l'influence des syndicats professionnels locaux." *Anthropology of Food*, 4. Online. https://journals.openedition.org/aof/286.
L'Union girondine des vins de Bordeaux. 1993. Bordeaux: Feìdeìration des syndicats des grands vins de Bordeaux. 884.
Maleysson, Fabienne. 1995. "Vins français, qualité en péril." *Que Choisir?* 321: 15–16.
Montaigne, Éric. 1994. "Bilan et perspectives de la réforme de l'OCM." *La Journée Vinicole* 18: 908.
ONIVINS. 2003. *Vins de France: une culture en mouvement: Les vins français face à la concurrence internationale.* Colloque organisé à l'Assemblée nationale. Paris: L'Office national interprofessionnel des vins.
———. 1998. *ONIVINS infos*, 52. Paris: L'Office national interprofessionnel des vins.
———. 1995. *ONIVINS infos*, 28. Paris: L'Office national interprofessionnel des vins.
———. 1994. "GATT: les dispositions applicables au secteur viticole." *Progrès Agricole et Viticole* 111 (8): 175–78. Paris: L'Office national interprofessionnel des vins.
———. 1993. *Réactualisation des statistiques sur la filière viti-vinicole.* Paris: L'Office national interprofessionnel des vins.
Parasecoli, Fabio. 2017. *Knowing Where It Comes From: Labeling Traditional Foods to Compete in a Global Market.* Iowa City: University of Iowa Press.

Paris Court of Appeal, 1st room A., December 15, 1993.

Petiteville, Franck. 2006. *La politiques internationale de l'Union européenne*. Paris: Presses de Sciences Po.

Pijassou, René. 1970. "La viticulture bordelaise dans la seconde moitié du XVIIe siècle." In *Vignobles et vins d'Aquitaine*. Bordeaux: FHSO.

Princen, Sebastiaan. 2004. "EC Compliance with WTO Law: The Interplay of Law and Politics." *European Journal of International Law* 15 (3): 555–74.

Réjalot, Michel. 2007. *Les logiques du château: filière et modèle vitivinicole à Bordeaux, 1980–2003*. Talence: Presses Universitaires de Bordeaux.

Regulation (EEC) No 817/70 of the Council of April 28, 1970 Laying Down Special Provisions Relating to Quality Wines Produced in Specified Regions, Official Journal of the European Communities, No. L 99/20, May 05, 1970, pp. 252–57.

Revel, Jacques. 1996. "Présentation & Micro-analyse et construction du social." In *Jeux d'échelles: La micro-analyse à l'expérience*, directed by Jacques Revel. Paris: Gallimard/Le Seuil.

———. 1989. "L'histoire au ras du sol." Forwards to Levi, Giovanni, *Le pouvoir au village*. Paris: Gallimard.

Roger, Antoine. 2010. "Constructions savantes et légitimation des politiques européennes. La circulation des savoirs sur la vigne et le vin." *Revue française de science politique* 60 (6): 1091–113.

Scheinmann, Lawrence. 1966. "Some Preliminary Notes on Bureaucratic Relationships in the European Economic Community." *International Organization* 20 (4): 750–73.

Scheinmann, Lawrence, and Feld Werner. 1972. "The European Economic Community and National Civil Servants of the Member States." *International Organization* 26 (1): 121–35.

Smith, Andy, Jacques de Maillard, and Olivier Costa. 2007. *Vin et politique: Bordeaux, la France, la mondialisation*. Paris: Presses de Sciences Po.

Smith, James. 2001. "Cultural Aspects of Europeanization: The Case of the Scottish Office." *Public Administration* 79 (1): 147–65.

Stanziani, Alessandro. 2003a. "Action économique et contentieux judiciaires: Le cas du plâtrage de vin en France, 1851–1905." *Genèses* 50 (1): 71–90.

———. 2003b. "La falsification du vin en France, 1880–1905: un cas de fraude agro-alimentaire." *Revue d'histoire moderne et contemporaine* 50 (2): 154–86.

Stevens, Anne. 2002. "Europeanisation and the Administration of the EU: A Comparative Perspective." *Queens Papers on Europeanisation*, (4): 11.

Treaty establishing the European Economic Community. 1957. Rome.

Trondal, Jarle. 2003. "Political Dynamics of the Parallel Administration of the European Commission." CES Working Papers, p. 37.

Unwin, Tim. 1996. *Wine and the Vine: A Historical Geography of Viticulture and the Wine Trade*. New York: Routledge.

8 Modern resilience of Georgian wine

Geographical indications and international exposure

Anastasiya Shtaltovna and Hart N. Feuer

Introduction

Georgia, a former Soviet republic located on the southern slopes of the Great Caucasian Mountains, is reputed to be the cradle of wine (Anderson 2013; McGovern 2003, 2009). It has experienced around 8,000 vintages, boasts at least 525 endemic varieties of grapes, features distinctive production techniques centered on the use of clay vessels, *qvevri*, and has an enviable reputation for hospitality involving lavish and lengthy feasts (Anderson 2013). Wine plays an important role in everyday life in modern Georgia but is also embedded deeply in the physical landscape and economic livelihoods. The relatively small land area of Georgia is home to archaeologically important agroecological landscapes, a distinct culinary culture and grape varieties largely unknown outside of its historic trading region.

One iconic image of traditional Georgian wine production (i.e., the Kakhetian method) is the use of *qvevri* for long-term fermentation. *Qvevri* are large clay vessels that are typically buried in the ground up to their neck and in which wine is fermented and stored (Glonti 2010; Barisashvili 2011; Field Notes 2015). During the fermentation process, which occurs naturally and without additives of any type, the buried *qvevri* remain sealed with a ceramic lid. The wine is left to mature for up to six months before being opened, then the unfiltered but clear wine is ready to be tapped or bottled (World Bank 2015). This millennia-old technique is still used all over Georgia. Wines were usually not filtered, and in fact, the ferment purposefully included not only the grape juice but also pulp, including seeds and skins.

Many occasions such as everyday meetings, communal work and collective decision-making are accompanied with wine, such that its production, distribution and consumption are structural anchors in Georgian identity (Manning 2012; Van Assche, Shtaltovna and Hornidge 2016). Georgia has a distinctive tradition of wine consumption and gastronomy, coupled with the continuing practice of sourcing wine through social networks and informal, often-illicit production. The persistence of these dense social networks around wine, and a continued practice of household production in rural areas, means that the search for new vintages and determination of quality are resolved socially rather than by experts, governments or institutes.

The persistent distinctiveness of this tradition suggests that Georgian wine is not a case of delayed modernization but of an intentional continuation of wine traditions over generations, in spite of many technological and institutional alternatives. This orientation, however, cannot be taken for granted; rather, it should be seen as a testament to Georgia's cultural resilience over the past 8,000 years. Not only has Georgia survived the conventional sort of social, military and economic upheavals, but it has also seen its wine tradition often become an explicit target of wartime aggression, sabotage and foreign co-optation. This chapter will review some of these challenges. The most recent threat to this tradition, however, has emerged in the context of Georgia's integration into international capitalist markets after the dissolution of the Soviet Union.

Counteracting the specific threats to Georgia's wine sector faced by neoliberal market forces, including over-commodification, exploitation of labor, deterioration of agrobiodiversity and hegemonic trade standards, is arguably a challenge that Georgians have not encountered historically. However, Georgia's recent post-Soviet autonomy in the 1990s coincided with the reengagement of world trade and intellectual property authorities on a measure that seemingly provided the tools for managing capitalism's impact on heritage agri-food systems, namely geographical indication (GI). GI refers to products with specific characteristics, qualities or reputations resulting from their geographical origin. This differentiates products based on unique local features, history or distinctive characteristics linked to natural and human factors, such as soil, climate, local know-how and traditions (Food and Agriculture Organization of the United Nations 2018). At face value, this measure provides a convenient and seemingly calibrated package of countermeasures that would not only help ensure the continuation of Georgia's distinctive wine traditions but also provide opportunities to attract global attention to Georgia's wine while creating high-value exports for rural development. Fitting its narrative of resilience, the government of Georgia moved fast to embrace GI laws in the 1990s despite the major headwinds caused by two civil wars and a rural sector that was still reeling from its transition out of the Soviet economic system. Even before being able to join the World Trade Organization (WTO) and benefit from GI protections under the WTO Agreement on Trade-Related Aspects of Intellectual Property Rights (TRIPS) in 2000, Georgia established national GI laws, inked a number of bilateral GI treaties and joined the Madrid Protocol for the international registration of marks (1998). In many senses, Georgia decisively set the stage for its full participation in the international mechanisms of GI.

And yet Georgia's pathway to adopting, and adapting, GI is faced with structural contradictions. On the one hand, GI is designed to acknowledge and protect the combination of traditional know-how and its attendant agroecological systems. On the other hand, the framework for terroir under which GI was created in Western Europe is at odds with some of the basic modes of recognizing quality and consumption in Georgia. To begin with, as will be clarified in more detail, quality is fluidly created among the dense social networks for wine and, consequently, wine quality determination is less fixed and under the domain of experts. Indeed, the constant consumption and social networking around wine in Georgia creates a dynamic and multimodal space for assigning quality that might be dampened by reified models of quality recognition such as GI. While wine engagement, as expressed by personal relationships to growers, sourcing wine for special events and recursive learning about varieties and regions, is not automatically eclipsed by more static certification mechanisms such as GI, over the long term, this personal engagement may lose its cultural relevance to label-reading consumers. One more visible concern here is that the European organizing principle of terroir pushes Georgia to represent its wines principally on a geographical basis, with "product diversity" arising from the combinations of soil, water, climate and local know-how. However, the vernacular association of wine in Georgia most often begins with wines categorized according to variety (e.g., Saperavi, Mtsvani, Rkatsiteli, etc.), which in fact represents significant diversity in a country with more than five hundred endemic wine grape varieties (Charkviani 1962; Glonti 2010). While this aspect can be resolved in various ways in the codes of practice written for each GI, it is a truism that geographical area, rather than variety, always sits at the top of the organizational hierarchy of GI. Privileging a foreign, rather than local quality determination mechanism, even when restricted to a parallel system for export, can have unintended impacts on domestic consumption. Conserving the vernacular modality of wine organization and the important wine social networks presents a new challenge in establishing GI in Georgia. In this chapter, we highlight that one of the first signs that GI is disrupting traditions in transitional Georgia is the growing disparity between the elite and working-class support for modern wine protection tools such as GI.

This chapter reviews the optimistic narrative of GI and several critical perspectives about the universality of GI. First, we explore the sociopolitical factors in a potentially contentious country, whose agri-food tradition precedes and diverges from that of the hegemonic Western European countries who established the GI system. We will discuss what makes Georgian wine different from Western European wines. Second, we will discuss the role of the government and elite in Georgia. Being top-down or overprotective of Georgian farmers, they have played a crucial role in protecting the diversity and uniqueness of Georgian wine and farmers for thousands of years. Third, we discuss how appropriately GI captures the agri-food culture to which it is applied. Currently, GI helps to face various challenges brought by capitalism, such as (a) the growing demand from the Russian and Chinese markets for cheap wines; (b) the establishment of a positive Georgian wine reputation outside of Eastern Europe; and (c) the fight against counterfeits of Georgian wine. And fourth, we question whether the trend of GI is to become increasingly monolithic in its structuring of the wine sector in Georgia or whether it can sustainably become a discrete and parallel framework focused on export wine and tourism. Georgia's historical legacy of resilience is being tested once again, this time to determine to what extent it can emancipate its wine sector from the threats of global capital or whether it will be compelled to compromises in its historic mission.

A history of resilience in Georgia's wine diversity and quality. Why Georgian wine is different?

Georgia has always been a meaningful contributor to wine culture throughout history (Kharbedia 2015; National Wine Agency 2014). Indeed, as the cradle of wine, it is almost a truism to state that wine spread throughout the world from the territory that is now Georgia (McGovern 2003). However, while the basic evidence affirming modern-day Georgia as the origin of viticulture is growing in certainty, archaeological evidence is also increasingly implicating Georgia as the origin of advanced and differentiated viniculture (Kharbedia 2015). Evidence in archaeological records places Georgia as possibly the first region of the world to produce wine (Chkhartishvili and Darchiashvili 1980; Glonti 2010; Maghradze et al. 2016). Wine production goes back as far as 6,000 BCE (McGovern et al. 2017; Kharaishvili, Chavleishvili and Natsvaladze 2014; Gomarteli 2017). Excavations have revealed not only grape seeds but also all of the accoutrement of wine production: clay vessels (*qvevri*), special knives for trimming vines and traces of tartaric (wine) acid (Chkhartishvili and Darchiashvili 1980; Glonti 2010). The purpose here is not to suggest that Georgia lays claim to higher forms of authenticity but rather to acknowledge in advance that protective measures for wine cultural practices, such as GI, are imported largely from the French or Italian experience and that it is fair to probe the hegemonic dimension of this transfer.

Besides its archaeological relevance, Georgia has also played a quiet, yet important, role in the institutionalization of wine culture. For example, Georgian wine, which precedes and likely plays a role in the spread of the cult of Dionysus, is acknowledged in the Homerian epic around 700 BCE (National Wine Agency 2014). And later, following the country's relatively early conversion to Christianity in the 4th century CE, both wine and the vine acquired political and religious meanings. Perhaps as a result of equating wine with the blood of Christ (a belief held in common within the whole Christian world), Georgia developed some of the earliest religious production traditions, in this case an integration of *Church, vineyard* and *irrigation channels*. This triad was often self-contained, a small administrative unit with theocratic powers. Through this religious legitimation and accumulation of power, viticulture played a part in efforts to preserve political and administrative unity and became one of the pillars for the country's government. Indeed, even historical documentation of pre-Christian Georgia inform us that the planting and irrigation of vineyards were symbols of the establishment of a state (Kharbedia 2015; National Wine Agency 2014).

This association of viticulture and state power in Georgia meant that wine played a role in official history – in diplomacy and trade but also conquest

(Kharbedia 2015). As a relatively small nation, distinct in religion from its non-Christian neighbors, Georgia was often a target, with its vineyards considered symbolic of the polity. After adopting Christianity, Georgia suffered numerous invasions at the hands of the Mongols, Persians and Turks, with the marauding armies promptly setting out to destroy the vineyards (Kharbedia 2015; Field Notes 2013, 2015). In the informal conception of Georgian history, vulnerability and invasion feature prominently but so does the stubbornness of the Georgians in their painstaking replanting of the vineyards.

Wine plays an important role in Georgian everyday life. Simple hospitality includes wine and toasting, but special guests and events are often celebrated with a *supra*, a more formalized gathering with lots of food and wine and many toasts organized by a *tamada* or toastmaster (Manning 2012). The *tamada* tells people when to drink, an action that has to happen collectively and promptly. The *supra*, in different guises, marks many events structuring the life of an individual and community, and thus it reproduces the embeddedness of wine in community life. Dense social networks around wine, and a continued practice of household production in rural areas, means that searching for wine and establishing quality are also often resolved communally. This extends also to the many monasteries, which produced and consumed wine in large quantities. The wine cellar of the Nekresi Monastery, for example, occupied around 200 m^2 and contained five wine presses capable of crushing 10 Mt of grapes at a time (Kharbedia 2015).

The 19th century was one of the most important periods in the history for Georgian winemaking, as it marks early initiatives to modernize wine production and the beginning of Georgia's role as a satellite wine producer. Under the Russian protectorate, prince Chavchavadze traveled frequently to Europe and invited French wine experts to Georgia (for example, the French vine expert Massono). Already in the 1830s, Chavchavadze started integrating some Western European grape varietals and methods of wine production, such as the targeted aging of wines, use of new tools for the care of vineyards and early forms of marketing, such as bottling (Interview with Batiashvili A., wine technologist, Chavchavadze vinery, Tzinandali/Kakheti, 2013; Charkviani 1962). Probably the most significant difference was the use of oak barrels (as opposed to ceramic amphora/*qvevri*) to mature the wine. But these changes were far from monolithic, and many of the changes ultimately adopted widely were reflections on, rather than direct transfers from, European winemaking. For example, Chavchavadze, among others, helped established more widely accepted, marquee varieties Rkatsiteli and Saperavi as the main wine grapes for the well-known Kakhetian region (Charkviani 1962), much like how chardonnay or cabernet sauvignon came to dominate many European regions.

In general, from the 19th century, Georgia's outward orientation was bracketed by national habits; it only cautiously adopted or integrated foreign winemaking practices for trade or elites while often holding onto indigenous practices for most domestic consumption. Even as wood barrels became more common, the traditional Georgian wine production (i.e., the Kakhetian method) using *qvevri* vessels, has been repeatedly reappreciated for its utility, by both domestic and international observers. The earthenware walls of *qvevri*, made of clay and rich with mineral material, contribute to a particular flavor profile (Glonti and Glonti 2013). The unfiltered yet clear wine that emerges has been acknowledged as efficient and gustatorily beneficial. Famous winemakers of that time, such as G. Lants and M. Ballas, after having observed the Kakhetian winemaking method for years, wrote that it has "an undoubted advantage . . . to the European barrels and vessels" and noted its authenticity and uniqueness (Lants 1846; Ballas 1877; Glonti and Glonti 2013). The adaptations to the Kakhetian method have been incremental. The European methods introduced in the 19th century encouraged separating the grape juice quickly from the rest, an activity that was partially integrated into the Kakhetian method in the form of a secondary fermentation. Although the prevalence of these new methods increased, contemporary Georgians consume primarily bulk wine fermented in the historic manner (Field Notes; Interviews 2015).

These adaptations helped create alternative products for external markets, such as Russian elites. For this segment, Georgian wine simply needed to be refined sufficiently for it to offer a comparable product to European wines. Thanks to the efforts of Prince Chavchavadze and those of the Russian viceroy, Mikheil Vorontsov, exported Georgian wine came close to matching the respectability and patrimony of European wine products by the end of the 1830s; in this period, the first European-style wine cellars were established in Georgia (Kharbedia 2015). Thus, Georgian wine has historically developed parallel products for elites or export, a characteristic that bodes well for its adoption of international GI products. Consequently, the latest uptake into world trade (i.e., China, Russia, the United States, Germany, etc.) is not unprecedented. Here already, however, the question emerges of how much agency Georgia has had in its international engagement and to what extent these engagements have appreciably impacted domestic winemaking and consumption. Although these questions cannot be answered directly, evidence emerged through a sequence of two major shocks to winemaking over the proceeding century: the outbreak of phylloxera and the advent of the Soviet Union.

The first shock, caused by the phylloxera mite that had destroyed many European vineyards in the 1850s and 1860s (Interview with Meghradze D., Institute of Horticulture and Viticulture, Tbilisi 2013; Chkhartishvili and Darchiashvili 1980), encouraged institutional changes to Georgian vineyard management from end of 1880s. As the Europeans did, Georgians also had to experiment with grafting endemic varieties onto the stems of American wild grape species (Charkviani 1962), and in general the plague sped up the development of scientific wine development in tsarist Russia, which oversaw Georgia. The government established the first institute for viticulture in Georgia in the late 19th century. Despite this event, Georgia did not move to high production varieties. In fact, largely following historical trends, Georgia doubled down on its endemic varieties. After these events in the 19th century, studies were compared the qualities and requirements of the local grape varieties, and importantly, a rigorous mapping of the varieties was conducted (Van Assche, Shtaltovna and Hornidge 2016). Hundreds of local varieties were discovered, more than most other world regions combined. There is now broad agreement that the number of discrete endemic varieties in Georgia is between five hundred and six hundred (Charkviani 1962; Glonti 2010). With international cooperation, including most prominently a group of Russian scientists, an eight-volume series was compiled covering 525 varieties of wine, their production methods and ecosystems. Here we see that, as in its long history, when faced with an external shock, Georgia doggedly worked to reestablish and maintain the diversity of Georgian grape varieties. To some degree, the documentation and reemphasis on traditional varieties represents a precursor to GI, but without the marketing and intellectual property focus of the modern legislative framework.

The next major shock, which dragged on for roughly seventy years after Russia's annexation of Georgia, arose from the way in which it was included in the Soviet Union. Perhaps fortunately for Georgia, the country was appointed as a wine-producing republic, but the transition from small peasant farming to collective production, combined with the strong role of Russian knowledge and trade institutions, radically reshaped the wine landscape. Viticulture became one of the main branches of agriculture of Soviet Georgia and, in many of its regions, the leading industry (Agricultural encyclopedia 1949; Van Assche, Shtaltovna and Hornidge 2016). Georgian wines quickly gained popularity in the Union of Soviet Socialist Republics (USSR). Furthermore, the products of Georgian winemakers at this time even received recognition in European countries, where they repeatedly won prizes at international exhibitions. It was in the Soviet era that the production of such popular brands as Kindzmarauli and Akhasheni began (Dernyatin 2005).

The production of grape wines was taken care of by Samtrest, established in 1924. Samtrest was a state department of vines and wine and it regulated the issues of viticulture and winemaking in Georgia. Samtrest took over the entire winemaking industry of Soviet Georgia and became the owner of all the existing wine cellars and later came to dominate all wine production in Georgia (Kharbedia 2015). The wine production promoted in the Soviet era consisted of several lines: table wines, dessert wines, sparkling wines (*shampanskie vina*) and fizzy wines (*shypuchie vina*). The highest quality, and most characteristic for the wine areas, were the premium wines (*marochniye*).

In the 1930s and 1940s, Georgia was able to overtly reproduce many endemic grape varieties and integrate those into a diverse array of sixty different types of wine. Over time, however, rationalization largely pushed the reproduction of wine types and varieties underground, with only twelve of the sixty wine types ultimately being based on local technologies (Kharbedia 2015). Some of these were based on the *qvevri* method but were somewhat scaled up or industrialized (Amerine and Joslyn 1970). In general, the production of wine skewed toward the broad tastes of consumers in Russia and the rest of the USSR. In particular, Russian demand in the early 20th century was for large quantities of low-quality, semi-sweet mostly red wines (Anderson 2013; Kharbedia 2014,

2015). The main task of Georgian winemakers was to fulfill the state plan. Thus, production oriented toward quantity over quality. This also meant that, over time, grape varieties with higher output began to crowd out even the few varieties openly produced for the Soviet wine plan (Apziauri 2017). Ironically, after investing heavily in large-scale wine production, Georgia suffered a major blow from the anti-alcohol campaign (*sukhoy zakon*) launched in 1985 by Mikhail Gorbachev (Dernyatin 2005). Vineyards were cut down, and wineries were closed. This law plunged the Georgian winemaking industry into a crisis which lasted into the 1990s (Kharbedia 2015). And so here again, Georgia had to find a way to recover from external shocks while rebuilding its wine heritage.

Post-Soviet developments of Georgian resilience and wine

Although the dissolution of the Soviet Union can be understood, in the longer term, as a release valve for the Georgian wine sector, the political and economic violence of the post-Soviet era brought about new challenges. To date, the Georgian government is still struggling to achieve basic policy adjustments, but this hurdle has not apparently slowed down the public drive to reestablish an independent wine sector (Shtaltovna 2016). The rural sector, in contrast, is still recovering from this period and struggling to embrace the myriad differences faced in a liberal market environment. The first major challenge arose through post-Soviet land privatization because it provided small plots to many families, including those with no prior agricultural experience (USAID 2011; Shtaltovna 2016). This step complicated the rationalization of agriculture, in the sense of consolidation, mechanization and professionalization, but it did quickly help to reinvigorate the previously suppressed diversity of wine and production styles.

Here, it is helpful to draw a picture of the post-Soviet farmer. Until independence, Georgian farmers had grown accustomed to top-down state directives. A farmer had for generations been understood as a character who lacked agency and who apparently does not have the means and expertise, agriculturally or managerially, to play the role ascribed to them by elite planners and international standards agencies. Land privatization, ironically, materialized this view: almost everybody received land and thus everyone became a farmer, including those in professions wholly unrelated to agriculture, such as medical doctors, teachers and accountants. Thus, the monolithic notion of "farmer" did not really match the intentions of agricultural or subsistence producers that emerged in Georgia and most of the former Soviet countries after independence (Shtaltovna 2016). Being fragmented in this way and emerging from Soviet centralization, "farmers" expected the government to act in a protective, guiding manner. When combined with the deficit in expertise and infrastructure, it is not surprising that post-Soviet farmers were not entrepreneurial or quick to engage (critically) with new policy experiments, such as GI. So the expectation that farmers, and by extension winemakers, at the small and medium scale of production – who represent a large swathe of agrobiodiversity in Georgia – would appreciate the marketing potential and intellectual property protection afforded by GI is not a given.

One particular challenge in the context of the post-Soviet mentality is that farmers are averse to entering into cooperatives or cooperative-like structures given the fresh trauma of forced collectivization. GI certification necessarily requires some formalized structure, usually an association, to establish a consensus about the basic standards for a production area. An organizational structure based on mutual interdependence might appear suspiciously like a new form of collectivization, particularly as the government has been encouraging it in a top-down manner since the 1990s. Furthermore, such a structure would require producers to fix shared standards, which they are culturally unaccustomed to doing in the dynamic cultural economy of Georgian wine. These factors explain why the government's attempts to push producer ownership of GI associations has not been successful in the preceding two decades and why local ownership does not obviate concerns about the capacity of producer associations to protect the unique production and agrobiodiversity of Georgian wine and to establish democratic and representative governing structures.

Hoping to later align rural production realities to policy innovations, the Georgian government quickly moved to equate its indigenous tradition of recognizing the geographical place of origin to the dominant appellation modality present in the international wine market (Sakpatenti 2010). Although Georgians use a multimodal method of determining wine quality, with geographical origin as a subsidiary consideration, territorial demarcations have thus come to increasingly play a more prominent role in differentiating Georgian wine for external markets (Interview with Kasradze 2018; Anderson 2013). For example, during the difficult transition period in the early 1990s, when Georgia was engaged in two civil wars, illicit use of Georgian territorial markers frequently took place in former Soviet countries. Many former consumers of Georgian wine from the Soviet period were being misled by counterfeits to purchase low-quality imitations, which negatively impacted Georgia's reputation (Sakpatenti 2010; Interview with Kasradze 2018). The government and Samtrest therefore undertook urgent measures to shore up intellectual property protections for regional appellations. In 1998, a law on Vines and Wine was signed, and Georgia joined the Madrid Protocol on the protection of international marks. In 1999, the Law on Appellations of Origin of Goods and Geographical Indication in Georgia (hereinafter, "the Law") was approved. The Law regulates the registration, protection and use of appellations of origin and GIs of goods (and services). The Law also set out the basic procedures for wine production, regulated wine quality and protected the market against counterfeit and low-quality products (Kemashvili 2012). As part of the Law, eighteen GI wines were registered (Sakpatenti 2010). By 2000, Georgia joined the WTO and became a signatory of TRIPS, which provided a platform for strong international protections (Interview with Kasradze 2018). This was followed by a presidential decree (2002) on the protection of GI as an urgent matter in order to avoid misappropriation within the country and to more adequately prepare a basis for legal claims at the international level (Interview with Kasradze 2018). This decree aligned the rules for the identification, registration, use and control of GI wines and brandies with TRIPS and set the groundwork for Georgia to join the Lisbon Agreement for the Protection of Appellations of Origin and their International Registration in 2004 (Kemashvili 2012).

Even as Georgia rushed to resolve counterfeiting and quality issues, a major shock lay in wait when, in 2006, Russia embargoed Georgian wines. The embargo was lifted in 2013 (Eurasianet 2013). The closing of Georgia's biggest market led to panic among Georgian winemakers, but such a sharp external shock was not new for the Georgian wine sector. Indeed, it harkened back to Gorbachev's anti-alcohol campaigns in the mid 1980s. Ultimately, many producers took this in stride, even considering it an opportunity to escape from the centuries of hegemonic Russian influence over Georgia's wine sector (Kharbedia 2015). Even in the toughest times, vineyards were seen as a good investment, as something that would always bounce back because of its deep local roots and a belief in the eternal foreign curiosity about Georgian wine (Van Assche, Shtaltovna and Hornidge 2016; Interviews 2013). In fact, when Georgian winemakers were pressed to export the volume previously earmarked for the Russian market to other countries, it led to increased awareness of Georgia outside its historical trade area in Eastern Europe and the Caucuses. But awareness of Georgia abroad does not quickly, or even necessarily, translate into motivation to cultivate granular knowledge of wine territoriality among wine connoisseurs.

From post-Soviet to international: accommodating external wine standards in Georgia

As Georgia began the slow and fragmented process of translating its wine diversity to foreign markets, a number of marketing pathways opened up.[1] In terms of protecting and promoting Georgia's range of wine offerings, some pathways (such as GI) were understood to be superior to trademarked or bulk wine sales, but it was still questionable whether the type of GI practiced for wine in Europe (sui generis) could fairly capture the unique agroecological diversity and quality associations in Georgia. Domestic sensibilities about how to organize wine quality, unsurprisingly, share many considerations with the European outlook, but they also differ in important ways from methods developed in Italy, Portugal and France (Van Assche, Shtaltovna and Hornidge 2016; Interview with Zibiashvili 2018). What is consistent between Georgian and European wine is the emphasis on agroecological considerations and know-how, but as we will discuss shortly, this emphasis can create blinders to other equally important considerations in the Georgian context.

People in the Kakheti region were called the wine people by communities in the mountains and further west. Knowledge of the terroir, of small differences tied to the village location and production method but also differences between parcels and plots, in terms of exposure, soil, drainage, slope and microclimate, were discussed by everyday people and related to the type and quality of wine (Interview with Nikolaishvili N., Tsinandali, a small winemaker, Kakheti region 2013). Kakheti became the epicenter of wine production and prestige by combining this know-how with the agroecological characteristics of the region (Chkhartishvili and Darchiashvili 1980; Amerine and Joslyn 1970). Kakheti is a mostly flat region in the wide Alazania valley that encompasses a few smaller hilly areas, and 67 percent of all vineyards of Georgia are in Kakheti (Kharaishvili, Chavleishvili and Natsvaladze 2014). The climate is dry, with spring rains, and features diverse sources of water. Local variations in soil and topography create conditions for diverse wine production (Sakpatenti 2010; Glonti 2010). A common meeting place would be the basement in old houses where the *qvevri* were dug in, sometimes bottles kept (although bottles were the exception). This was a place with social significance, dominated by men, who would come together to help each other in winemaking, discuss quality, plan for the rest of the year and drink heartily. Even in the Soviet era, people remember that it was considered shameful to treat guests with wine purchased from a store (Apziauri 2017). Producers would always keep the better wines for guests and themselves; wine of poorer quality would be sold on the market or on the street (Mukhranov 2017). In general, Georgian experts and winemakers agree that, as in times past, peasants and small entrepreneurs were the primary stewards of Georgian wine tradition during the Soviet era. Through their often-unauthorized informal production, they helped protect many of the unique varieties of wine and the integrity of unique production systems (Apziauri 2017; Mukhranov 2017), which are now reemerging in the post-Soviet era. Notwithstanding the importance of terroir, as understood in its European origin, there are key differences in the organization of wine and assessment of quality that are less suited to the legislative framework of GI.

The most obvious distinction is that wine grape varietal places an outsized role in a country with considerable diversity. The estimated five hundred to six hundred varieties in Georgia (Charkviani 1962; Glonti 2010), with a population of around 3.7 million (excluding Abkhazia and South Ossetia) far outpaces the next competitor, France, which has (in the most generous estimate) 273 varieties for a population of 67 million.[2] In searching for wine (through one's networks or in shops), variety is often the first point of reference. In the sui generis model of GI, in contrast, geography is so important that often varietals are not easily found on labels or, in some cases, not listed at all. Because labeling does not have to follow any specific convention, there is nothing stopping Georgians from prominently displaying varietal, but over the long term, the geographical origin may, for reasons of trade standards or marketing, displace this preexisting hierarchy. There is optimism that this problem can be mostly obviated by placing GI as a secondary point of qualification and avoiding varietal specification in GI codes of practice.

In the assessment of quality, Georgians continue to uphold a strong tradition of multimodal quality determination that is situated in the dense social networks of wine. While this does not, at face value, contradict sui generis GI standards for qualification, it questions the suitability of the formalized institutional structures typical of GI certification. The durability of these informal social networks around wine, and the persistence of household production in rural areas, suggests that quality determination by experts or the government might be considered unwelcome in the domestic market. The resilience of the Georgian wine sector is built on its defense of cultural capital against assertive external demarcation initiatives, and this resistance is likely even higher now due to post-Soviet suspicion of grand top-down restructuring schemes. The concerns of producers are also not entirely obviated by the founding of more local producers associations, which are typical for GI certifications worldwide, because such associations do not have a natural geographical basis in rural Georgia and would marginalize an important stakeholder, namely consumers. This is an important consideration in a society where significant wine consumption forms the backbone of social relations, and certain wines are developed purposefully for certain events, ceremonies or times of the year (Manning 2012; Van Assche, Shtaltovna and Hornidge 2016). A wider stakeholder group than is typical of GI producer associations would be necessary to overcome this institutional difference.

A final difference that complicates the adaptation of sui generis GI certification mechanisms for Georgia is the need to raise the stature of the *qvevri* style of fermentation, which plays a significant role in Georgia. While the use of *qvevri* can be inscribed in respective codes of practice for GI areas, and thereby gain transparency, the stark nonconformity of *qvevri* with the hegemonic understanding of wine fermentation vessels in the world is perhaps worthy of more attention. The current state of expertise and consumption expectations surrounding high-quality wines assumes the use of wood barrels or stainless steel vats. Georgia, as a small country with limited cultural capital at its disposal, may struggle to educate world consumers about its unique form of fermentation and aging, not to mention the wide range of unfamiliar wine varietals. The support of multilateral agencies such as the United Nations Educational, Scientific and Cultural Organization (UNESCO), and the nascent fascination by wine sommelier and travel guides in unique wine, can hopefully help Georgia overcome its deficit in marketing capital. Georgia is also aggressively pursuing a policy of trade diversification and promotion of its wine in world markets that is beginning to break into new territory.

Wine in contemporary Georgia: resilience through trade diversification

A number of events around the 2010s mark Georgia's increasing integration into the international community, especially the European Union. In 2013, the Georgian *qvevri* method was approved by UNESCO as a world Intangible Cultural Heritage. According to UNESCO,

> the tradition plays a vital role in everyday life and celebrations, and forms an inseparable part of the cultural identity of Georgian communities, with wine and vines frequently evoked in Georgian oral traditions and songs.

> Knowledge of this heritage is passed down by families, neighbors and friends, all of whom join in the communal harvesting and wine-making activities.
>
> (World Bank 2015)

This was further buttressed by definitive archaeological findings confirming Georgia's 8,000-year pedigree of wine (Glonti 2010; Kharaishvili, Chavleishvili and Natsvaladze 2014; Maghradze et al. 2016; McGovern et al. 2017). Not only was Georgian wine increasingly visible and being taken seriously, but its accessibility was increasing through Georgia's pivot toward liberalism and the European Union. Around 2015, in a stream of EU integration processes, including a major free trade agreement, a number of joint Georgia-EU documents concerning agriculture and rural development were signed that directly affected the wine sector (Ministry of Agriculture of Georgia 2015; Association Agenda between the European Union and Georgia 2014). By tying together policy agendas promoting wine diversification (particularly GI) and trade diversification (i.e., wider trade networks), Georgia furthered its domestic initiative to create value around its wide range of wine patrimony.

Efforts to diversify trade have been helped by Georgia's new recognition as a producer of distinctive, quality wines. For Georgian wine exports, 2017 was a record year, with data showing that 76.7 million bottles of wine were exported to fifty-three countries, representing an increase of 54 percent compared to 2016 exports (National Wine Agency 2017a). In 2018, Georgia continued to sell a lot of wine. Georgia exported about 86.2 million bottles of wine to fifty-three countries in 2018, which is a record high in the last thirty years (Agenda.ge 2019). The top importers are still former Soviet countries, including Russia (33 million bottles), Ukraine (6 million bottles), Kazakhstan (2.3 million bottles), and Poland (2.1 million bottles). However, China now imports 4.1 million bottles from Georgia, with Western Europe and the Americas receiving around 1 million bottles. Export growth is strongest in the European Union, the United States, Asia and former Soviet countries (National Wine Agency 2017b). This is made possible by improving grape harvests and better wine promotion. "By the end of 2018, about 100 million bottles of wine are expected to be exported," according to Levan Dolmazashviliashvili, Minister of Agriculture (National Wine Agency 2017a).

Exporting wine outside of Georgia's historical trading area has required continual engagement to create awareness, achieve standards and negotiate trade relations. As discussed above, the immediate post-Soviet period included a flurry of activity in this respect, including the signing of treaties on intellectual property and joining the WTO. Georgia has continued in this trend by continuously adding bilateral agreements that address certain concerns (counterfeit, standards) and/or open trade routes. For example, in 2007, intergovernmental agreements on the mutual protection of the geographical indications of wines, spirits and mineral waters between Georgia and a number of countries went into force, including the United States, China and several former Soviet countries

(Sakpatenti 2018). In 2012, the EU-Georgia Agreement on Mutual Recognition of Geographical Indications of Agricultural Products and Foodstuffs came into force. Particularly toward China, there continues to be a growing trend of exporting Georgian wine and spirits, which increasingly skews toward higher-quality GI wines (Egutia 2013; Interview with Dolmazashvili 2013).

For a small, lower-middle-income country like Georgia, increasing global recognition and international agreements and improving export values are important macro-level achievements. While this success cannot be underestimated, the fact that top-down governmental initiatives have been the driving force raises concerns about the extent to which rural producers are benefiting, particularly those who have continued to steward Georgia's diverse wine tradition. Given the weak post-Soviet rural sector, the Georgian government arguably stepped in to proactively position Georgian wine as best as possible in the global capitalist economy (Kemashvili 2012).

Benevolent authoritarianism in Georgian wine institutions

Since the dissolution of the Soviet Union, the Georgian government has been the leading actor in deploying GI and other wine promotion policies. This is not to suggest that public sector involvement in GI promotion is unusual but instead that the roles ascribed to producers and professional associations in the Western European model have, in Georgia, largely been driven directly by the state. Samtresti, together with the government, encourages farmers and other producers to take over initiatives, such as GI producer groups, by suggesting that they "re-register the GIs to the farmers' associations and production cooperatives," but "[the stakeholders] have not been very active in this process" (Interview with Kasradze 2018). In the meantime, the state sees no other choice but to push on undemocratically by institutionalizing the future structures of the wine trade and hoping that the wine sector matures into the projected structures (Interview with Kasradze 2018). Since many of these activities appear inevitable given worldwide trends, such as joining the WTO, establishing sui generis GI regulations, fostering trade abroad and promoting wine heritage at UNESCO, the government's actions appear to be benevolent and proactive. The Georgian government believes that GI can be a good tool for building awareness and improving the reception and value of Georgian wines, and it aggressively promotes GI in its national development strategy (Ministry of Agriculture 2015). The European Union has also recognized this, granting €1.5 million (US$1.7 million) to Georgia to improve its legislation and quality control and to build capacity along the entire wine value chain (Interview with Kasradze 2018; Sakpatenti 2010). But by adopting what appears to be structurally preordained and by limiting the future room to maneuver for the wine sector, the government has also denied agency in this respect to many of the stakeholders who have stewarded the Georgian wine tradition over generations.

This potentiality, however, is at best abstract to observers of the wine sector. Historically, elites have always engaged with the wine trade internationally, while the domestic sector followed a separate, vernacular course. The national intellectual property center of Georgia (Sakpatenti), which was established to promote GIs among wine producers, is nominally meant to cooperate with wine producers, exporters and business associations on raising awareness among different parties. While Sakpatenti interacts well the Samtresti and various relevant line ministries, it is poorly connected to small-scale producers (Egutia 2013). Here the task of adopting a foreign concept such as GI and strategically positioning Georgian wine for international audiences is understood as the purview of the government – as long as they do not impinge on the local idiosyncratic modes of production and consumption. One illustration of this is the government's pivot away from the hegemonic orientation of terroir based on geographical location to one based on variety:

> Our products are different from wines supplied to Russia from other countries: we do not offer international and widely known varieties: Chardonnay, Cabernet Sauvignon or Merlot. In Soviet times, these wines were supplied to Russia mainly from Moldova, and in recent years, Moldovan wines compete with Chile, Argentina, Australia and other countries. From Georgia, Russia receives an original, unique product: Saperavi and Rkatsiteli, these are varieties that are not found anywhere else. Kindzmarauli and Khvanchkara are recognizable in Russia and have found their former consumer.
>
> – Levan Dolmazashviliashvili, director of the Georgian National Wine Agency
>
> (RFE 2014)

Although Georgia shares the European principles of terroir, it situates geographical origin lower on the hierarchy. Indeed, many shops in the Georgian capital of Tbilisi shelve wine based on variety rather than origin, reflecting the different constellation of quality valuation (Field Notes 2015). Defending this variation of terroir has significant implications for potential international trade. The Georgian government is concerned with the vulnerability of small grape farmers whose diverse range of production has historically been a dynamic asset in the Georgian wine landscape but who are now exposed to international market forces. One major worry is associated with the conversion of some vineyards to international varieties in order to cater to foreign tastes. One response has been to embrace a mechanism built into GI by encouraging a gradual move from protected geographical indication (PGI), which allows for the use of grapes from distant areas, to protected designation of origin (PDO), which would require wine producers to more exclusively seek out and compensate farmers for locally produced grapes (Interview with Kasradze 2018). Although this biases the hierarchy of quality toward geographical origin, rather than variety, the potential of higher economic returns associated with the maintenance of hyper-local wine heritage is encouraging a growing number of small and medium-size wine-producing enterprises to spring up in rural communities, creating employment opportunities, supporting traditional winemaking practices and breathing new life into rural economies (Kemashvili 2012).

Here the Georgian government has, in comparison to many countries at similar levels of development, obviously moved relatively fast to institutionalize GI for wine from a technical standpoint and more generally to integrate Georgian wine into global trade. However, whether the top-down or state-driven promotion of Georgian wine will ultimately help to protect the diversity and uniqueness of Georgian wine depends on its eventual uptake by farmers and everyday consumers. The initial "benevolent authoritarian" stage of GI, which was ostensibly initiated in the 1990s to expeditiously begin using international tools for protecting Georgian intellectual property, began at a time when the wine industry was reeling from perestroika and ongoing civil wars, not to mention the restructuring and privatization of farming. Farmers and factories were considered by the government to be vulnerable and in need of protection, although they could perhaps not grasp the scale of the pending problem at that time (Interview with Kasradze 2018). In this sense, the embrace of GI in the face of the still-abstract threat of commodification of Georgian wine culture can be considered a progressive move by the government. Whether the government can incrementally democratize this policy shift remains to be seen.

Conclusion

Georgian wine can be characterized by its resilience. Despite appropriation by cults and religions, destruction by invaders and civil strife, reorganization by the Soviet Union and transformation by capitalism, vineyards have always found ways to survive, maintain agrobiodiversity and uphold the unique cultural embeddedness of wine in Georgian society. Explaining this phenomenon with a broad brush is impossible, but clearly, one of the secrets of this Georgian resilience is the alignment of elite patronage – royal or governmental – in promoting national wine culture and persistent peasant support for wine in everyday life. This combination of top-down and bottom-up resilience is now on display again in Georgia's attempt to engage with the international wine market.

To some degree, Georgia's quick uptake of intellectual property laws, such as GI, which might help secure its wine heritage in the face of the commoditization pressures of the capitalist market, suggests that this resilience has taken on a new, modern form that is adapted to the challenges ahead. Some of these "new" challenges have familiar echoes, such as the growing demand from the Russian (and now Chinese) markets for inexpensive but fabled wines and the fight against the counterfeiting of Georgian wine abroad. But other challenges, such as establishing the reputation of Georgian wine outside of its home territory in the former Soviet republics, require a completely new positioning and branding of Georgian wines and wine heritage.

In this, Georgia is faced with a contradictory strategy in the framework of GI. On the one hand, GI is theoretically designed to acknowledge and protect the combination of traditional know-how and its attendant agroecological systems. On the other hand, the framework for terroir under which GI was created in Western Europe diverges in slight but important ways from some of the basic modes of organizing and recognizing quality in Georgia. This difference clusters around three main areas, none of which preclude the utility of GI but instead highlight its awkward fit in Georgia and the potential long-term disruption of the hegemonic construction of quality emanating from Europe. The first difference is that the vernacular association of wine in Georgia most often begins with wines categorized according to the variety (e.g., Saperavi, Mtsvani, Ojaleshi, etc.), which in fact represents significant diversity in a country with more than five hundred endemic wine grape varieties (Charkviani 1962; Glonti 2010). This complicates the establishment of geographically defined producer organizations and the marketing of Georgian wine on the international market. The second difference is that the constant consumption and social networking around wine in Georgia creates a dynamic and multimodal space for assigning quality that is antithetical to quality determination by experts and government. While it is not uncommon to find preexisting quality determination schemes coexisting with GI, assertive intervention in the cultural domain of wine has a long history of being resisted in Georgia. The third difference is that the promotion of *qvevri* wines from Georgia, while technically possible to render transparent in codes of practice, will require a challenge to the hegemonic expectations of wine as fermented in wood barrels or stainless steel. Even with the recognition of *qvevri* winemaking as Intangible World Heritage by UNESCO, raising awareness about the range of quality considerations is a major undertaking for a small country with little cultural capital.

The question now is whether the dynamism and uniqueness of Georgian wine culture can be fairly captured by relatively statist, standards-focused policy mechanisms like GI or whether GI can be deployed, in a sustainable fashion, in parallel to preexisting wine recognition systems. Although GI is unlikely to become the monolithic organizing principle for wine in Georgia, it is fair to question whether GI may displace or disrupt some of the informal reproduction of wine culture in the long run. Because the European concept of terroir differs in important ways from Georgian principles of organizing and valuing wine, as described in the section about accommodating external wine standards, tension may emerge if domestic-focused production and consumption are fixed rather than allowed to be fluidly allocated in the dense social network around wine. The history of the Georgian Prince Chavchavadze, who is recognized for having built a discrete sector of winemaking for export, suggests that international quality attribution mechanisms such as GI can remain undisruptive if contained at the elite levels or if applied only as an expedient tool for communicating quality markers of Georgian wine.

The risks of Georgia's proactive and renewed public engagement with the international wine trade should be viewed in light of its historic capacity for positioning its wine despite adverse geopolitical conditions. Most relevant here is Russia's recent dominance over Georgian wine export, which began at the end of the 19th century, continued on through the Soviet era and persisted thereafter. To some degree, this hegemonic influence kept Georgian wine sequestered from other regions of the world. Since the dissolution of the Soviet Union, however, Georgia has been trying to reestablish sovereignty over its wine sector through domestic autonomy and diversification in export partnerships. These efforts came into sharp relief in 2006, when ongoing geopolitical tensions resulted in a Russian embargo on Georgian goods, including wine. By pivoting toward other regions, namely the European Union, China and other former Soviet Republics, Georgia hopes to draw more attention to its storied wine heritage and reduce its dependency on traditional export partners. Doing so, however, has also required Georgia to embark on new ways to render its products and quality transparent and accessible to markets with different tastes, expectations and standards. The rapid embrace of GI by the government, despite the slow uptake and ownership by producers, is therefore indicative of its policy first, democratize later approach, which is anchored in Georgia's manifestly successful history of benevolent authoritarianism.

Meanwhile, the domestic market remains vibrant and, in many ways, unaffected by the shifting patterns of Georgia's engagement in export markets. To outside observers, Georgia may seem like a pre-industrial Spain or Italy, in that everyone still drinks large quantities of peasant-made wine, but that the change to more commoditized wine consumption seems inevitable.

This chapter presents both historical and current evidence to suggest that Georgia is likely to maintain this consumption pattern as it is (a) inimitable to the unique wine production system centered on *qvevri*-fermented wine and the use of endemic varieties, (b) anchored in the practice of sourcing and valuing wine through social networks and (c) grounded in a long history of employing informal, often-illicit production to survive external shocks. Georgians faced such crossroads throughout history and more recently in the 19th century with the outbreak of phylloxera, after which it expeditiously returned to production and consumption of local wines varieties. The country faced this crossroads again in a protracted manner during the Soviet era, when local idiosyncratic production was preserved by peasants in an unauthorized manner. As a small country, maintaining a parallel engagement with international trade and standards while providing domestic space and encouragement for local production represents the efficacious past combination. Whether the Georgian government can build on this historical success by creatively adapting GI to the wine sector or by keeping it aloof from domestic production and consumption patterns has yet to be seen.

Notes

1 The following two-paragraph section is adapted from the author's previous shared publication, Van Assche, Shtaltovna and Hornidge 2016.

2 According to ONIVIN (Office national interprofessionnel des vins), FranceAgriMer Stats 2010.

Literature

Agenda.ge. 2019. "Georgia Exports 86.2 mln Bottles of Wine in 2018, Record High of Last 30 Years." Accessed January 25, 2019. http://agenda.ge/en/news/2019/23.

Agricultural Encyclopedia. 1949. *Viticulture*. Edited by P. Lobanov and others. Vol. 1. 3rd ed., revised. Moscow: State Publishing House of Agricultural Literature.

Amerine, Maynard A., and Maynard Alexander Joslyn. 1970. *Table Wines: The Technology of Their Production*. Berkeley: University of California Press.

Anderson, Kym. 2013. "Is Georgia the Next 'New' Wine-Exporting Country?" *Journal of Wine Economics* 8 (1): 1–28.

Apziauri, Sofo. 2017. "How Georgian Wine Has Survived?" Accessed October 20, 2018. https://jam-news.net/how-georgian-wine-has-survived.

Association Agenda Between European Union and Georgia. 2014. Accessed March 15, 2016. https://eeas.europa.eu/diplomatic-network/black-sea-synergy/6913/association-agenda-between-georgia-and-eu_en.

Ballas, M. 1877. "Essay on Winemaking in Russia (The Caucasus and Crimea)." *Historical and Statistics Collection on Winemaking*. Volume #1 (Original in Russian).

Barisashvili, Giorgi. 2011. *Making Wine in Qvevri – A Unique Georgian Tradition*. Tbilisi: Elkana.

Charkviani, K. 1962. *Winemaking and Its Place in the Economy of Soviet Georgia* (Origin in Russian), 1–206. Tbilisi: *Zarya Vostoka*.

Chkhartishvili, N., and R. Darchiashvili. 1980. *Viticulture in Georgia*, 1–37. Tbilisi: Ministry of Agriculture of GSSR, Sabchato Sakartvelo.

Dernyatin, Andrey. 2005. "Numerically-Catalogued Wine: Wine and Traditions of Winemaking – The Main Strategic Resource of Georgia that is Poor in Mineral Resources (Origin in Russian)." Accessed October 15, 2018. www.sostav.ru/news/2005/11/08/45/.

Egutia, Ekaterine. 2013. "Marketing and Protecting Geographical Indications of Georgia Abroad." Presentation at the World Intellectual Property Organization, Bangkok. Accessed October 10, 2018, www.wipo.int/edocs/mdocs/geoind/en/wipo_geo_bkk_13/wipo_geo_bkk_13_18.pdf.

Eurasianet. 2013. "Georgia-Russia: A Diplomacy of Wine and Water." Accessed August 15, 2015. www.eurasianet.org/node/66499.

FAO. 2018. *Strengthening Sustainable Food Systems Through Geographical Indications*. Rome: FAO Investment Centre. Food and Agricultural Organization of the United Nations.

Glonti, Teimuraz. 2010. *Traditional Technologies and History of Georgian Wine*. Tbilisi: Institute of Horticulture, Viticulture and Oenology.

Glonti, Teimuraz, and Zurab Glonti. 2013. *The Remarkable Qvevri Wine: Why Is the Qvevri Wine Better Than the One Made Without It*. Tbilisi, Georgia: The *Qvevri* Foundation.

Gomarteli, Nina. 2017. "Success Year of Georgian Wine." *Caucasus Business Week*. Accessed October 15, 2018. http://cbw.ge/wine/successful-year-of-georgian-wine/.

Kemashvili, Elene. 2012. "Uncorking Georgia's Winemaking Potential." *World Intellectual Property Organization Magazine*. Accessed October 29, 2018. www.wipo.int/wipo_magazine/en/2012/02/article_0003.html.

Kharaishvili, E., M. Chavleishvili, and M. Natsvaladze. 2014. "Trends and Prospects for the Development of Georgian Wine Market." *World Academy of Science, International Journal of Economics and Management Engineering* 8 (10): 3259–63.

Kharbedia, Malkhaz. 2015. "Georgia: The History of Georgian Wine." *Wine Club*. Accessed October 25, 2018. http://en.vinoge.com/history/history-georgian-wine.

———. 2014. "Nostalgic Image of Georgian Wines of Soviet Era Should Come to an End in Russia." *Wine Club*. Accessed October 25, 2018. http://en.vinoge.com/articles/malkhaz-kharbedia-nostalgic-image-georgian-wines-soviet-era-should-come-end-russia.

Lants, G. I. 1846. "The State of Winemaking and Treatment of Grapes in Kakheti." *Newspaper 'Kavkaz'* 10 (Original in Russian).

Maghradze, David, Giorgi Samanishvili, Levan Mekhuzla, Irma Mdinaradze, George Tevzadze, Andro Aslanishvili, Paata Chavchanidze, David Lordkipanidze, Mindia Jalabadze, Eliso Kvavadze, Nana Rusishvili, Eldar Nadiradze, Gvantsa Archvadze, Patrick McGovern, Patrice This, Roberto Bacilieri, Osvaldo Failla, Gabriele Cola, Luigi Mariani, Nathan Wales, M. Gilbert, P. Thomas, Laurent Bouby, Tina Kazeli, Levan Ujmajuridze, Stephen Batiuk, Andrew Graham, Lika Megrelidze, Tamar Bagratia, and Levan Davitashvili. 2016. "Grape and Wine Culture in Georgia, the South Caucasus." In *BIO Web of Conferences* 7, p. 03027.

Manning, Paul. 2012. *Semiotics of Drink and Drinking*. London: Bloomsbury.

McGovern, Patrick E. 2009. *Uncorking the Past: The Quest for Wine, Beer, and Other Alcoholic Beverages*. Berkeley: University of California Press.

———. 2003. *Ancient Wine: The Search for the Origins of Viticulture*. Princeton, NJ: Princeton University Press.

McGovern, Patrick E., Mindia Jalabadze, Stephen Batiuk, Michael P. Callahan, Karen E. Smith, Gretchen R. Hall, Eliso Kvavadze, David Maghradze, Nana Rusishvili, Laurent Bouby, and Osvaldo Failla. 2017. "Early Neolithic Wine of Georgia in the South Caucasus." *Proceedings of the National Academy of Sciences* 114 (48): E10309–18.

Ministry of Agriculture of Georgia. 2015. *Strategy for Agricultural Development in Georgia 2015–2020*, 2–12. Tbilisi.

Mukhranov, Aleksey. 2017. "Georgian Home Made Wine (Original in Russian)." Accessed October 20, 2018. http://travelgeorgia.ru/831/.

National Wine Agency. 2017a. "11 Month Data of Wine and Alcohol Beverage Export 2017." Accessed October 15, 2018. http://georgianwine.gov.ge/En/News/2511.

———. 2017b. *Report of Activities of the National Wine Agency in 2017*. Tbilisi. Accessed October 15, 2018. http://georgianwine.gov.ge/En/Files/Download/1041.

———. 2014. *Georgian Wine Culture*. Tbilisi.

Radio Free Europe (RFE). 2014. "40 Million Bottles Are the Joys of Export (Original in Russian)." Accessed October 27, 2018. www.svoboda.org/a/25256557.html.

Sakpatenti. 2018. *Georgian Geographical Indications Protected Abroad*. Tbilisi. Accessed October 29, 2018. www.sakpatenti.gov.ge/en/page/113/.

———. 2010. *Appellations of Origin of Georgian Wine*. Tbilisi.

Shtaltovna, Anastasiya. 2016. "Development of Extension Services in Post-Soviet and Post-Conflict Georgia." In *Extension in Post-conflict States*, edited by P. McNamara and A. Moore. Boston, MA: CABI.

USAID. 2011. *Economic Prosperity Initiative: Sector Assessment Report*. Final. Prepared by Deloitte consulting LLP. USAID/Caucasus, pp. 1–233.

Van Assche, Kristof, Anastasiya Shtaltovna, and Anna-Katharina Hornidge. 2016. "Local Knowledge and Expert Knowledge in Rural Transition: Georgian Wine Production." In *Agricultural Knowledge and Knowledge Systems in Post-Soviet Societies*, edited by Anna-Katharina Hornidge, Anastasiya Shtaltovna and Conrad Schetter, 223–50, 15. Bern, Switzerland: Peter Lang.

World Bank. 2015. "What's in a *Qvevri*? Georgia's Intangible Cultural Heritage." *Feature Story*. Accessed October 15, 2018. www.worldbank.org/en/news/feature/2015/07/20/whats-in-a-*qvevri*-georgias-intangible-cultural-heritage.

Part IV

Cases from the Americas

9 The multilevel, multi-actor and multifunctional system of geographical indications in Brazil

Paulo Niederle, John Wilkinson and Gilberto Mascarenhas

Introduction

GI constitutes a strategy for distinguishing goods based on their sociocultural embeddedness in the territory in which they are produced (Cerdan 2013). While there is little controversy with regard to this general understanding, debates over the implementation of this mechanism have proliferated in recent decades and unmasked a number of contradictions: Is it a non-tariff barrier to global trade or a support policy for territorial development? Is it a mechanism for protecting cultural goods or an instrument for imposing new technical standards on food goods (Josling 2006)? Even in the European context, where GI was initially created by states to protect certain territories against the usurpation of their historical reputation, GI has become much more than a public instrument for the protection of common goods (Belletti, Marescotti and Touzard 2017). In different contexts, GIs can be the result of private actor strategies to differentiate product quality drawing on the reputation of a specific region or the promotion of organizational and technological innovation. They can also be promoted as exclusive "club goods" and may be appropriated by corporate economic actors (Thiedig and Sylvander 2000).

A similar understanding of these multiple functions of GIs can also be found in countries which have availed themselves of this instrument more recently after adhering to the World Trade Organization (WTO)Agreement on Trade-Related Aspects of Intellectual Property Rights (TRIPS), signed in 1994, which includes a provision for GI (Food and Agriculture Organization of the United Nations2018; Marie-Vivien and Biénabe 2017). In each country, however, GIs have to be adjusted to different cultural, economic, political and institutional realities, leading to a proliferation of different systems for the recognition and protection of products of origin. As is analyzed in other chapters of this book, some countries approximate to the European model based on public and collective property rights and territorial development, while others promote private systems of market regulation, where GIs are recognized and regulated as trademarks (Sekine, this volume; Renard and Domínguez Arista, this volume).

Some countries have adopted a third trajectory creating sui generis systems, where GIs are distinctive from trademarks or any other well-institutionalized property rights system. Brazil fits closest into this last category, developing a flexible and hybrid institutional arrangement which involves the articulation of public and private actors at various institutional levels (territorial, state and federal). In addition to different interests for investing in a GI, the various actors often have divergent understandings of its nature, whether public, collective or private. The GIs in Brazil are therefore profoundly shaped by the "political compromises" concluded in each sector or territory, leading locally to quite specific institutional and organizational models, despite common national and international regulations (Wilkinson, Cerdan and Dorigon 2017).

This chapter analyzes how this multilevel, multi-actor and multifunctional system operates in three sectors: wine, cheese and coffee, which together constitute fourteen of the fifty-eight GIs registered in Brazil as of July 2018. The chapter explores the relation between the different results produced by these GIs and the institutional and organizational flexibility which characterizes GI in Brazil. At the same time, we show how this flexibility is responsible for the adaptation of GIs to different realities but equally how it favors the development of misunderstandings on the nature of GIs and the competences of the different organizations involved in their governance. The price of flexibility, therefore, is an institutional instability which limits the potential of GIs both in their promotion and regulation of new markets and in their promotion of territorial development.

Research questions and methodology

Several studies have revealed the heterogeneity of GI systems throughout the world, which engage different networks of actors, governance mechanisms and support policies and which have different impacts on territorial dynamics (Food and Agriculture Organization of the United Nations 2009; Barjolle et al. 2017). They have also shown that GIs are tools that can be adapted to a wide range of goals: achieving trade protection for specific markets, creating territorial oligopolies, ensuring monopoly rents and raising quality standards – and the consequent exclusion of less competitive firms and lower income consumers (Food and Agriculture Organization of the United Nations2018; Cerdan 2013). A number of authors have also suggested that, through the hybridization of competitive agri-food strategies, GIs have aligned market models which were previously considered to be antagonistic (Sekine and Bonanno 2017). At the same time that GIs promote the valorization of territories and their specific identities, they can also create barriers to innovations which place these identities at long-term risk. On the other hand, they may also help producers to react to the loss of market competitiveness by providing rewards for choosing a non-industrial pathway that leverages the unique links between these products and their territories.

Since registering the first GI in Brazil in 2002, various studies have attempted to measure their results. This task has been particularly difficult given the wide range of sectors which have been the object of GI strategies: foodstuffs, artisan crafts, industrial products and even services, which are relatively innovative in Brazilian Industrial Property Law (LPI no. 9.279/1996), since this sector is not mentioned in the TRIPS Agreement. These studies identified four major "functions" of GIs for achieving development goals: (1) value creation and market access; (2) innovation and the adoption of differentiated production practices; (3) the valorization of natural and intangible heritage; (4) collective involvement and incentives for cooperation, the so-called project effect (Mascarenhas and Wilkinson 2014).

Some of these studies also highlight the prevalence of two competing conceptions of 'GI for Development' in Brazil. One conception focuses on the recognition and patrimonialization of cultural heritage, while the other accentuates market access and the modernization of technical processes. In the wine sector, for instance, Niederle and Vitrolles (2010) have identified a "double process of institutionalization," converging toward two extremes. On one side, GIs are seen as an institutional innovation that stimulates technological modernization and enables a catching-up process for the lead firms, while scarcely affecting local inequalities (and sometimes amplifying them). On the other side, they are conceived as part of an effort to reify traditional practices of production while creating possibilities for including new actors in value chains and stimulating the creation of more plural governance structures in these territories. Below, we will explore to what extent these sorts of antagonisms might be generalized to other sectors and territories.

Aiming to generate results covering multiple sectors or regions, we developed a collective research project from 2013 to 2016 with financing from the Brazilian National Council for Scientific and Technological Development(CNPq). The project included researchers from different universities across the country and from the National Public Agricultural Research Corporation (EMBRAPA), together with the participation of the National Institute for Industrial Property (INPI) and the Ministry of Agriculture (MAPA). The central focus of the research was to investigate the principal results of GIs from the perspective of development and to identify the respective factors explaining the different results by sector and territory. However, as the research advanced, it became clear that efforts at quantification encountered the difficulty of isolating the specific impact of the GI initiative from other general processes affecting the sector or region, namely value creation, market access, the promotion of cultural heritage and technological change. The research was therefore redesigned to focus on the relation between the institutional frame of GIs and their different and even contradictory development dynamics.

Drawing on Ostrom's (1990) notion of a "nested institution," GI is understood as a complex result of diverse actors' strategies with their different cognitive frameworks and interests. To explore GIs within this optic, three different sectors were selected: wine, cheese and coffee. Each was analyzed along five dimensions: (5) international agreements on GIs; (2) GIs within the framework of Brazil's Intellectual Property Laws; (3) GI regulations for specific sectors; (4) related regulation on relevant product and service sectors; and (5) the GI codes of practice. The research objective here is to analyze how the actors adjusted their strategies to the multilevel institutional framework in terms of value creation, access to the market, technological innovation, the protection of the ecological heritage and the promotion of cultural goods and collective involvement. To achieve this, in addition to documentary research, more than a hundred interviews were carried out between 2014 and 2016 with farmers, entrepreneurs, researchers and public administrators. The results, which we present in what follows, show that, given the low level of institutional enforcement of international and domestic norms in Brazil, the GIs have been defined in line with the specific interests prevailing in each sector and territory. The resulting institutional incoherencies create obstacles for the consolidation of a market for GI products in Brazil and equally for using GIs as a public policy instrument of territorial development.

Wine, coffee and cheese: three paradigmatic examples

Wine

GI legislation has been decisive for anchoring definitions of quality to territorial criteria in the wine sector and aligning these with the differential protection within the TRIPS Agreement. In Brazil, the construction of GIs for wine was initially envisioned as a quality-oriented strategy to enhance the capacity of national enterprises facing competition from imported wines (Food and Agriculture Organization of the United Nations 2018), which represented 48 percent of the domestic market in 2002, when Vale dos Vinhedos, the first Brazilian GI, was registered. The GI initiatives privileged the Serra Gaúcha region, where 85 percent of national production is located and where five of the six GI wine areas are now registered (Vale dos Vinhedos, Pinto Bandeira, Monte Belo, Farroupilha and Altos Montes). The Serra Gaúcha is a historical region of European colonization, with Italian immigrant culture having a role as important as the grapes themselves for the reputation of wine quality.

In this region, researchers from EMBRAPA were the first to argue that a GI could be an alternative strategy of market differentiation for winemakers. They proposed that the quality of the local wines be promoted by incorporating technical prescriptions in the GI code of practices and that, at the same time, the image of the product be connected to the cultural symbols of the Italian immigrant communities. Using their apparently disinterested position, these researchers engaged other actors and assumed the position of "macro-actor" in a new territorial network created to initiate the GI project (Callon 1986). In this network, they established themselves as "bridges" or "nodes" connecting producer associations and organizations involved with the laws and policies on intellectual property at the national and international levels (Niederle and Vitrolles 2010). Through their control of the negotiations, or processes of "translation" in actor-network language, between different social worlds, these actors were able to adjust the emerging institutional framework to their own interests and normative understandings. This involved the incorporation of foreign concepts, such as the French notion of terroir, which was adapted to the realities of the "new world," where, as one researcher expressed it, "innovation will always be a more active factor."

Given that the majority of these researchers were trained in agricultural sciences, it is not surprising that a technological approach became predominant in all the GI wine projects in the Serra Gaúcha. Aligning also with the market interests of the producer associations, this perspective created pressure to incorporate various technologies and practices that are common worldwide, such as vertical systems of grape cultivation and the substitution of hybrid varieties by European ones, into the GIs' code of practices. While this points to a standardization of production and processing practices, the codes of practices importantly drew in local norms, such as the prohibition of irrigation and plastic crop coverings, which according to the local actors, are a threat to the link between the product and its terroir. To achieve these new quality requirements, however, the processors decided to invest in their own grape cultivation areas rather than encouraging existing farmers to transition. As a consequence, most of the technical choices adopted became associated with exclusionary politics. Similar to the cases of the Mexican tequila and mezcal analyzed, respectively, by Bowen and Zapata (2009) and Renard and Domínguez Arista (this volume), the "politics of the terroir" in the Brazilian wine sector has also resulted in the marginalization of small grape producers, who are (designed to be) unable to achieve the procedures and standards.

The GI strategy has, paradoxically, been singularly ineffective in its aim to increase the domestic share of the Brazilian wine market, with imported wines reaching 86 percent of sales in 2018. However, in comparison to other processors, those positioned in GI areas have not only been more resilient to the crisis generated by the escalation of imports but also found openings in new markets, where they are obtaining premium prices for their wines. These include specialized retail segments and high-gastronomy restaurants, in addition to local and direct markets associated mainly with wine tourism. Although the development of these markets was driven not only by the GI strategy, this tool has promoted technical and organizational innovations essential to changing both the taste and the image of these regional wines. This recently acquired reputation has even encouraged the creation of new wineries in the region.

These sectorial economic benefits have a spillover effect favoring all the local winemakers of the GI areas whether or not they follow the code of practices or use the GI label. Recognition of "origin," however it is achieved, is more important than a fledgling label, particularly in a country where most consumers are still unfamiliar with the concept of GI. This "free-rider" effect, which in one sense can be seen as a positive externality for the region, has led to the "desertion" of the GI by some actors who no longer want to pay the costs of maintaining the collective strategy. The Vale dos Vinhedos Association started in 1995 with only six members and currently has thirty-one, but only a dozen of these use the GI label on their products. Taking advantage of a controversial regulation and the fragility of control mechanisms, the others simply quote the name of the region on their products which has the same reputational effect (Food and Agriculture Organization of the United Nations 2018).

On the territorial level, the qualification of origin-specific wine production has also encouraged investments in other sectors, such as hotels, restaurants and locally made artisan products. Even though GIs focus on a single product, this positive externality shows how value can be created for a more extended territorial "basket of goods" (Mollard and Pecqueur 2007). On the other hand, the increasing valuation of local resources following these new economic investments has prohibitively raised the opportunity costs for many smaller producers and processors. A major concern of local actors is the replacement of farms with upper-class residential condos and hotels. The situation is particularly dire for grape producers not engaged in any strategy to produce high-value products. Faced with an increasing profit squeeze, they are encouraged sell their land and/or migrate. In this way, the continuation of the grape cultivation of all forms is threatened and the integrity of the landscape and the culture of the territory with it.

Coffee

Although the Vale dos Vinhedos GI and, more generally, the wine GIs have become the principal reference for the construction of GIs throughout the country, similar debates developed in the coffee sector and have evolved in parallel fashion. The Council of the Association of the Cerrado Coffee Producers (CACCER) applied for a GI for coffee produced in the Savannah region of the state of Minas Gerais even before the one presented to INPI by group from Vale dos Vinhedos. Indeed, it is worth speculating about whether this coffee GI had become the first to be approved and thereby set the precedent for the construction of GIs in Brazil. However, of primary interest here is to contrast the conceptualization and implementation of the coffee and the wine GIs. In contrast with the wine GIs we analyzed before, which are all located in the state of Rio Grande do Sul, the four coffee GIs occupy areas in three different states – Minas Gerais (Cerrado Mineiro and Serra da Mantiqueira), São Paulo (Franca) and Paraná (Norte Pioneiro). We are dealing, therefore, with a much more heterogeneous set of ecological, sociocultural, economic and political conditions that must be distilled in the respective GIs' codes of practice.

Another important difference between wine and coffee concerns the nature of the product itself and the role of Brazilian producers in the broader market. Brazil is the world's leading producer and exporter of coffee, accounting for 30 percent of world supply and 33 percent of the global coffee trade. However, most of Brazil's exports are commodity coffee, which are quoted on stock exchanges and subject to market price fluctuations. Rather than a reaction to the entrance of imported products as in the case of wine, the coffee GIs are boosting Brazil's production of specialty coffees, which are less vulnerable to price fluctuations and, at the same time, have higher prices in international and domestic markets (May, Mascarenhas and Potts 2004). According to Mascarenhas and Bernardes (2017), specialty coffees now account for 12 percent of the international coffee market and have obtained price premiums ranging from 30 percent to 40 percent.

Two further differences should also be mentioned. In the first, the development of strong agglomeration effects identified in the wine region (i.e., tourism gastronomy, complementary craft industries), were not present in the coffee GIs analyzed. Coffee beans are a raw material input and downstream sectors are generally located outside the regions. A second point relates to the nature of the actors mobilized around the GIs, with coffee generally playing a more decentralized and smaller role for researchers. In each region, different actors assume leading roles: MAPA in the GI Serra de Mantiqueira, SEBRAE (the National Body for the Promotion of Small and Medium Enterprises) in the GI Norte Pioneiro and the CACCER Association in the Cerrado Mineiro. The absence of a "macro-actor" unifying these GI coffee initiatives has meant that the construction of the codes of practice has been less cohesive than in the wine region. The similarities that do exist in the norms and standards for coffee have been influenced by the diffusion of CACCER's original model, which has created a pattern of path dependence comparable to the wine region (Niederle and Gelain 2013). The Brazilian Association for Specialty Coffees, however, has also had an indirect impact on GIs to the extent that it has similarly defined grades and standards for specialty coffees and thereby provoked a measure of institutional isomorphism (DiMaggio and Powell 1983).

Nevertheless, GIs in the Brazilian coffee sector are similar to those in the wine sector in at least four dimensions: (1) they are primarily seen as a strategy for product qualification, with a goal of reconciling technical innovation with the valorization of local cultural heritage; (2) price premiums and differential market access spill over into non-GI producers, corroborating the systemic problem of free riding in Brazil's GI scheme; (3) the GI potential is limited by poor consumer awareness, particularly the image of low-grade commodity coffee in global trade and unfamiliarity with GIs in the domestic market (Mascarenhas and Bernardes 2017); (4) technical requirements mostly follow the logic of "quality escalation" – a continuous increase in grades and standards – which leads to the exclusion of producers with poor resources or those not located in the best terroirs.

Cheese

The construction of GIs for cheese in Brazil shares important features with the controversies over artisan cheese from raw milk in the Brazilian market. While much less developed than the historical tradition of cheese production in Europe, or even in neighboring countries such as Argentina, Brazil does have one, with the state of Minas Gerais being responsible for some 50 percent of national production. Regions in this state, such as Serra do Canastra, Serro and Salitre have acquired national reputations for their artisan cheese production.

Despite its popularity, however, artisan cheese production has been threatened by federal regulations advocating pasteurization and requirements that cheese made with raw milk be subjected to long periods of maturation. In the case of cheese, GI strategies have not been adopted primarily to combat falsifications, displace imports or drive niche exports, but they have aimed to develop high-end marketing strategies which might offset the costs of adapting to the demands of federal legislation. This marketing strategy has been accompanied by vigorous social mobilizations challenging the industrial criteria for cheese production and has achieved important reformulations, such as the formal recognition of traditional cheese production.

In the state of Minas Gerais, the organization of the dairy sector was part of an international cooperation program signed between the state government and the French Ministry of Foreign Affairs in the early 1990s. This agreement focused on quality, traceability and coordination along the production chain and was led in Brazil by the affiliate of the French non-governmental organization (NGO) Agrifert, in partnership with state actors such as the Public Corporation of Rural Extension (EMATER) and the Minas Gerais Institute of Agriculture (IMA). In response to the restrictive legislation, the Minas Gerais state government sanctioned a law within the confines of the state boundaries allowing for the marketing of cheese from raw milk with less than sixty days maturation if it is produced in regions known for their traditions of artisan production. It was in this context of legal reform that Agrifert, which had taken the lead in mobilizing both cheese producers and the different state organizations, suggested the GI project. Here, Agrifert came up with the idea that premium prices associated with this mechanism would offset the extra costs of adaptation. The additional GI strategy required an even broader mobilization, leading Agrifert to engage academia in studies on the typicity of these regional cheeses and on the maturation process. Simultaneously, this served as a challenge to the federal regulations.

After a decade of mobilization, GIs were granted to Canastra and Serro in 2011 and 2012, respectively. In the period after recognition, Agrifert reduced its operations, and a change in local government further reduced public support for GI. These changes exposed the low level of autonomy achieved by the producers associations, although this has been partially offset by the greater involvement of SEBRAE, which has invested heavily in rural artisan activities. Although not with the same intensity as in the Vale dos Vinhedos and other wine-producing regions, the promotion of the cheese GIs also led to the development of other territorial goods, such as tourism and gastronomical events.

In the cheese sector, GI strategies coexist with other mechanisms for recognizing and valuing artisanal products. The practices and knowledges associated with traditional cheesemaking are recognized as immaterial heritages by organizations such as Slow Food International and the Institute for National Historical and Artistic Heritage (IPHAN). The support of these patrimony-oriented organizations for more strictly traditional practices, however, leads them to adopt a critical stance toward technical innovations, which have been included in many GI codes of practice.

Even as the GI cheese strategies in Minas Gerais have challenged federal regulations and promoted traditional young cheeses products, they have also shifted toward mature cheeses, which can be considered a general reaction to the popularity of special quality cheeses in Brazil (a "market justification" in the language of convention theory). In opposition to the sixty-day maturation period required by federal regulation, the Minas Gerais regulatory agency specifies only seventeen days of maturation for the Serro region and twenty-two days for Canastra. Only some production avails itself of the exemption from federal law to produce young cheeses from raw milk, while many others have embraced the more mature cheeses. These long-matured cheeses have been further stimulated by the active promotion of these cheeses by nationally prominent chefs. Key producers in the region, especially those who have been involved in the GI movement are now able to sell a proportion of their cheese for a substantial premium in boutiques and markets and through on-farm sales. However, the adoption of these standards poses challenges for the characterization of traditional artisan production practices and has become the object of criticisms.

The investments required by the GI codes of practice have proved viable for only a small number of producers. Hence, of some 1,800 cheese producers in the Canastra region, only twenty are members of the Producer Association, while in Serro the number is seventy-six out of a total of almost one thousand. The great majority of traditional hill farmers producing cheeses continue to survive in the informal economy (Wilkinson 2016). It is perhaps for such reasons that Slow Food International, which supports second generation, consumer-driven and less administratively rigid versions of products of origin, has developed a global campaign in favor of artisanal cheese production from raw milk. In addition, a National Working Group on traditional cheese production was created in 2012, which is similarly challenging federal regulations, and has come out in favor of a broad recognition of artisanal cheese production from raw milk that does not demand costly investments or a shift to the matured cheeses of the gourmet market.

In Table 9.1, we provide a summary comparison of the general characteristics of each of the three sectors studied:

Institutional flexibility and systemic instability

When GIs were incorporated into Brazilian legislation, the prevailing idea supported by both governmental and non-governmental actors was that the law should outline a minimal normative framework for GI recognition, thereby transferring to the sectors and territories the responsibility for negotiating most of the specific rules and standards, which would then be incorporated into their respective codes of practice. As a result, although the broader international and national contexts should not be disregarded (as we see clearly in the case of federal regulation of dairy products), the development implications of the various GIs depend to a great extent on the vested interests and conceptions prevalent at the local level. These determine whether GIs serve a strictly market logic

Table 9.1 General characteristics of the GI projects in each sector

	Wine GIs	*Coffee GIs*	*Cheese GIs*
Initial Objective	Product qualification to respond to loss of markets to imports.	Product qualification to access markets for specialty coffees.	Adaptation of products and processes to Brazilian sanitary legislation and access to high-end markets.
Social Network	Articulated by EMBRAPA (federal public sector). Important presence of producer associations.	Dispersed Networks. Different state and non-state actors depending on the individual project.	Initially, Agrifert and Minas Gerais state (EMATER and IMA). Later, involvement of SEBRAE (quasi-state) and Slow Food.
Sectorial Dynamic	Vertical integration to fulfill quality requirements. Market segmentation. Exclusion of small-scale producers.	Favor capitalized farmers in mechanized regions and farmers with good terroir and high altitude, both for specialty coffees.	Difficulty of incorporating small-scale farmers in function of sanitary requirements for quality.
Territorial Dynamic	GI products strongly associated with other strategies which together promote a territorial mix of products and services.	Difficulties in creating positive complementary economic externalities at the territorial level. But general improvement in reputation for quality.	GI products reveal potential for the promotion of other goods, but the impacts are still quite limited.
Code of Practices	Norms and standards favor technical criteria and are similar in all the projects under EMBRAPA's coordination.	Norms vary in accordance with the characteristics of each region, but all are influenced by the standards of the specialty coffee sector.	Norms vary in accordance with the dominant coordinating actor, initially Agrifert, then SEBRAE, then Slow Food, IPHAN, social movements. They oscillate between modern technical quality and the valorization of traditional practices.

Source: Authors.

that privileges technological modernization and economic efficiency, or a more civic strategy rooted in the valorization of collective resources.

The lack of common norms for the same product category or sector exemplifies this flexibility of institutional arrangements in Brazil. In Europe, a PDO wine must be produced entirely from grapes grown in the demarcated area (85 percent in the case of a PGI). In Brazil these percentages are not regulated, so an indications of provenance (IP) can be even stricter than a denomination of origin (DO). The location of production, distribution and packaging provides a further example. In Europe, a product can be registered as a PDO only if all the stages of production are carried out within the demarcated area. Any exceptions to this rule must be justified. This requirement does not apply in the case of a PGI, where only those activities defined as essential to the quality of the product must be carried out in the demarcated area. In Brazil, there is no common norm in this sense, and as a result, a DO may be allowed to carry out certain production stages outside the demarcated area. All these decisions depend on agreements reached among local actors.

This understanding of the Brazilian model therefore exhibits the contradictory institutional arrangements which have marked the recent development of GIs in a number of other countries, many of which are analyzed in different chapters of this book. To accommodate some of the contradictory expectations of GI, the Brazilian GI system is relatively malleable, and in this way, it can be compared to many other policies for rural development. Many of these policies also oscillate between support for the commodity export sector within a liberal agenda and for a selective protection of markets in which family farming is the predominant segment (Wilkinson, Cerdan and Dorigon 2017).

The organizational and institutional flexibility of the Brazilian GI system has thereby created a series of anomalous GI cases which are making the system unstable and limiting its capacity to serve as a catalyst for the expansion of markets and territorial development. This is particularly the case when one considers the fragility of GI control mechanisms and the nonexistence of sanctions to punish inappropriate uses. Opportunist behavior is thereby stimulated and, if unchecked, leads some actors who bear the costs of certification to abandon the GI. To the extent that producers in the demarcated area who do not follow the GI code of practice (not to mention those outside the area) use the same geographical reference on their labels, the GI loses its ability to generate collective commitment. In this situation, producers turn to third-party certification schemes, collective marks and other classification such as Slow Food. As we have seen in the cases of both the wines of the Vale dos Vinhedos and the coffees of the Cerrado Mineiro, the coexistence of different mechanisms for recognition has only amplified the confusion surrounding these products.

The coexistence of two types of GIs in Brazil – IP and DO – further contributes to the uncertainty in the GI system. Formally, each designates a specific type of property right, with the former recognizing the reputation of a specific territory for a particular good or service and the latter indicating that the distinctive qualities of the good in question are the result of natural or human factors associated with the territory in which it is produced. Strictly from this differentiation, an IP does not need to exhibit any distinctive quality, whereas a DO does not legally depend on an established reputation. In practice, however, these definitions have suffered reinterpretations in each sector and territory. In the case of the Vale dos Vinhedos wines, the IP codes of practice include quality criteria, which are not formally required according to Brazilian legislation, since these aspects are globally understood to be better served with a DO. Perhaps for this reason, the IP in the wine region was seen as merely a first step toward a DO. Because the IP does not expire upon acquisition of a DO, in the Vale dos Vinhedos, both IPs and DOs with different codes of practice coexist. Only a tacit agreement in place locally inhibits producers from using the obsolete but more flexible IP code of practice, which shares the same name as the DO.

The transfer of comprehensive normative competences to the sectorial and territorial level is seen by many as one of the principal advantages of the Brazilian GI system since it allows for the distinctiveness of each case to be taken into account. Within the networks promoting GIs, such flexibility can, however, lead to the production of heterogeneous norms, which calls in question the national integrity of the GI system. For this reason, the recent attempt to regulate GIs through the adoption of technical norms drawn up by the Brazilian Association of Technical Norms (ABNT) has provoked divergent responses. Some see this initiative as an opportunity to reduce the instability of the GI system by introducing recommendations on terminology, traceability and good management practices. Others see an unacceptable shift from decentralized public governance to private governance since the ABNT is a private organization, a member of the International Organization for Standardization (ISO).

In light of these problems, discussions have been underway since 2012 on the reorganization of the Brazilian GI system. Proposals for new legislation have been drafted and discussed at various levels but have so far not been taken up by the National Congress. Innumerable conflicts have blocked advances in this direction. The most important of these relates to the fact that GIs are associated with broader legislation on industrial property, which is separately dealing with a host of institutional anomalies unrelated to the protection of products of origin. Alterations to the GI system, thus, threaten to open the Pandora's box of the whole system of industrial property in Brazil. A second source of conflict relates to the specific nature of GIs and their uses. Discussions on the construction of a new GI system have unmasked at least two basic positions. The first emphasizes GIs as a mechanism for the defense and valorization of common goods which are to be used by the government to promote the collective forms of territorial development. The second instrumentalizes the logic of "private collective rights," with GI seen as a market instrument which should be managed by producer associations in line with their specific interests, promoting what is known in the economic literature as "club goods."

These controversies reflect the historical processes which led to the internalization of GIs in Brazil and to the evolving conceptions of (new) stakeholders.

Although the principal legal mandate arose from the TRIPS Agreement, the Law on Industrial Property (LPI n. 9.279/1996) was also greatly influenced by the Mercosur Protocol for the Harmonization of Norms on Intellectual Property (Decision CMC n. 8/1995), which did not exactly reproduce the same terms as the WTO. Brazil defined its two modalities of GIs (PI and DO) based on the Mercosur Protocol, which correspond neither to the terms of the TRIPS nor to the European system. In practice, over the last two decades, the formal agreements signed by Brazil have been less important in determining the implementation of GIs than the informal influence of the European and especially the French model, which is the common reference for researchers, producers and administrators alike. European research organizations, particularly the French Agricultural Research Centre for International Development (CIRAD), have played a decisive role in the diffusion of GIs not only in Brazil but throughout the world.

Over the last two years, the renewal of negotiations for a trade agreement between the Mercosur members and Europe has been marked by the presence of a new issue. The European negotiators have put pressure on South American governments to adopt more rigid GI regulations, particularly on the creation of mechanisms of external control similar to the certification systems prevailing in the European Union (Marie-Vivien et al. 2018). In Brazil today, there is wide agreement among all actors that new mechanisms of control over GIs are necessary and important. There is less agreement, however, on the nature of this control and who should assume responsibility. Private actors resist the idea of a system of control by third-party certification, which would increase their costs and result in greater levels of exclusion. Even though the idea of GIs as public goods managed by the government has little support among the majority of the producers involved, they consider that the government should assume control over these issues. Brazilian law, however, differently from the European system, treats GIs as private property, and the government is unwilling to incur further costs.

Conclusion

Economic strategies focusing on the recognition of products of origin have become an alternative for many producers and territories. GI is one of many strategies, which include the connection between producers and consumers by means of short value chains or other political and social initiatives engaged in the valorization of food cultures. GI is therefore considered to be part of a broader and more complex process of relocalization and patrimonialization of the food system. This basic understanding was present among all sectors and territories analyzed in this chapter.

GI, however, has also been adjusted to suit purposes that are not necessarily linked to an origin-based strategy, such as the introduction of technical innovations and stricter quality standards that may contribute to a decoupling of the product and its terroir. In this case, GIs become open to appropriation by industrial actors, who may view GI as an institutional tool that provides advantages over competitors.

The impacts of GIs on development are clearly differentiated from one sector and territory to another. This is partially associated with the nature of the product, particularly the degree to which value can be added within the territory or will be added during distant processing. This creates special problems for raw materials such as coffee beans. The stronger association of wines and cheeses with local services (mainly gastronomy and tourism) favors a dynamic of territorial spillover in these regions that is not readily available to primary producers. However, the difference in institutionalization between products such as coffee, cheese and wine is sometimes not as significant as the differences between territories that have recognized a GI for the same product. This suggests that the outcome of a GI highly depends on the way it is articulated and combined with other strategies. If GI is used only to support a policy of product differentiation that is not anchored in shared local knowledge and horizontal networks, it will probably have a limited effect on the valorization of other territorial goods.

In each of the cases analyzed here, the construction of the GI has proved to be an important occasion for initiating processes of organizational innovation. This "project effect" is also highlighted by Feuer (this volume): "An observation not lost on many academic and civil society commentators is that GIs are able to generate a level of enthusiasm and social mobilization that is unusual in the otherwise dry field of intellectual property rights." Indeed, in the cases we have studied, we have identified a dynamic network of government and private actors (producers, researchers, agricultural technicians, etc.), not to mention a global stakeholder group. This is a new model of collective engagement for a country such as Brazil, where sectors are typically defined by "verticalized" forms of governance. As a result, in all contexts, we identified a variety of changes in the distribution of political power both in the specific sector and the associated territory.

This chapter has argued that the contradictory effects of GIs are a consequence of the institutional flexibility of this instrument in Brazil. A degree of flexibility is inherent given that GIs are designed to recognize and highlight the particularities of a given territory. In the Brazilian case, however, this flexibility derives also from the low levels of normative enforcement and from the conscious decisions of the legislators to transfer the greater part of decision-making to the sectorial and territorial actors. The result has been the creation of GIs which are not based on a set of common principles and which have led to the creation of alternative regulations to fill the gaps in national legislation. These regulations, however, have not addressed central questions such as social exclusion and the lack of disciplinary control in the event of unacceptable usage. The anomalies that have resulted from the proliferation of relatively independent subsystems have created a situation of institutional instability that reduces the efficacy of GIs to regulate the market and promote territorial development.

Bibliography

Barjolle, Dominique, Xiomara Quiñones-Ruiz, Monique Bagal, and Hermann Comoé. 2017. "The Role of the State for Geographical Indications of Coffee: Case Studies from Colombia and Kenya." *World Development* 98: 105–19.

Belletti, Giovanni, Andrea Marescotti, and Jean-Marc Touzard. 2017. "Geographical Indications, Public Goods, and Sustainable Development: The Roles of Actors' Strategies and Public Policies." *World Development* 98: 45–57.

Bowen, Sarah, and Ana Valenzuela Zapata. 2009. "Geographical Indications, *Terroir*, and Socioeconomic and Ecological Sustainability: The Case of Tequila." *Journal of Rural Studies* 25 (1): 108–19.

Callon, Michael. 1986. "Eléments pour une sociologie de la traduction: la domestication des coquilles Saint-Jacques et des marins-pêcheurs dans la baie de Saint-Brieuc." *L'Anné Sociologique* 36: 169–208.

Cerdan, Claire. 2013. "Indicações Geográficas e estratégias de desenvolvimento territorial." In *Indicações Geográficas: qualidade e origem nos mercados alimentares*, edited by Paulo Niederle, 125–50. Porto Alegre: UFRGS.

DiMaggio, Paul, and Walter Powell. 1983. "The Iron Cage Revisited: Institutional Isomorphism and Collective Rationality in Organizational Fields." *American Sociological Review* 48 (2): 147–60.

Food and Agriculture Organization of the United Nations. 2018. *Strengthening Sustainable Food Systems Through Geographical Indications: An Analysis of Economic Impacts*. Rome: FAO.

———. 2009. *Linking People, Places and Products*. Rome: FAO.

Josling, Tim. 2006. "The War on Terroir: Geographical Indications as a Transatlantic Trade Conflict." *Journal of Agricultural Economics* 57 (3): 337–63.

Marie-Vivien, Delphine, Laurence Bérard, Jean-Pierre Boutonnet, and François Casabianca. 2018. "Are French Geographical Indications Losing Their Soul? Analyzing Recent Developments in the Governance of the Link to the Origin in France." *World Development* 98: 24–35.

Marie-Vivien, Delphin, and Estele Biénabe. 2017. "The Multifaceted Role of the State in the Protection of Geographical Indications: A Worldwide Review." *World Development* 98: 1–11.

Mascarenhas, Gilberto, and Ricardo Bernardes. 2017. "A (r)evolução Cafés: o Resgate da Qualidade a partir das Origens." In *O Sabor da Origem*, edited by John Wilkinson, Paulo Niederle and Gilberto Mascarenhas, 213–56. Porto Alegre: UFRGS.

Mascarenhas, Gilberto, and John Wilkinson. 2014. "Indicações geográficas em países em desenvolvimento: potencialidades e desafios." *Revista de Política Agrícola* 23 (2): 103–15.

May, Peter, Gilberto Mascarenhas, and Jason Potts. 2004. *Sustainable Coffee Trade: The Role of Coffee Contract*. Montreal: IISD.

Mollard, Amédée, and Bernard Pecqueur. 2007. "De l'hypothèse au modèle du panier de biens et de services. Histoire succincte d'une recherche." *Économie Rurale* 300: 110–14.

Niederle, Paulo, and Delphine Vitrolles. 2010. "Geographical indications and qualification in the Brazilian wine production." *Estudos Sociedade e* Agricultura 18(se): 5–55.

Niederle, Paulo, and Julia Gelain. 2013. "Geographical Indications in Brazilian Food Markets: Quality Conventions, Institutionalization and Path Dependence." *Journal of Rural Social Sciences* 28: 26–53.

Ostrom, Elinor. 1990. *Governing the Commons: The Evolution of Institutions for Collective Action*. Cambridge, UK: Cambridge University Press.

Sekine, Kae, and Alessandro Bonanno. 2017. "Geographical Indication and Resistance in Global Agri-food: The Case of Miso in Japan." In *Resistance to the Neoliberal Agri-Food Regime: A Critical Analysis*, edited by Alessandro Bonanno and Steven Wolf, 106–19. London: Routledge.

Thiedig, Frank, and Bertyl Sylvander. 2000. "Welcome to the Club? An Economical Approach to Geographical Indications in the European Union." *Agrarwirtschaft* 49 (12).

Wilkinson, John. 2016. "Contested Markets: An Overview." *Antropolítica* 41 (2): 25–45.

Wilkinson, John, Claire Cerdan, and Clovis Dorigon. 2017. "Geographical Indications and "Origin" Products in Brazil: The Interplay of Institutions and Networks." *World Development* 98: 82–92.

Wilkinson, John, Paulo Niederle, and Gilberto Mascarenhas, eds. 2016. *O Sabor da Origem*. Porto Alegre: UFRGS.

10 The geographical indication of mezcal in Mexico

A tool of exclusion for small producers

Marie-Christine Renard and David Rodolfo Domínguez Arista

Introduction

As one of the types of geographical indications (GIs), denominations of origin (DO) identify and protect products whose quality or characteristics derive *essentially* or *exclusively* from their geographical environment, including natural and human factors (WIPO 1958). DO is, therefore, a more strict form GI that is adopted by the twenty countries that are signatories of the Lisbon Agreement (Avelino 2006). Case studies allow us to highlight the potential economic advantages of these protection strategies. First, there is the prestige markup or a premium that customers pay for the quality differences and reputation associated with the DO. Second, there is a territorial markup resulting from the reduced supply that derives from production limited to a given territory. Finally, there is a markup associated with a closed market generated by the limited entry, which constituted a de facto monopoly (Linck 1999). Many cases have shown that when properly applied, DO strategies favor the local and regional rural development of disadvantaged areas (Bowen 2011). This is the result of their production model and/or the creation of synergies with other activities, such as tourism. One of the most frequently cited examples is that of French Comté cheese that by privileging livestock that produce less has engendered a sustainable rural development model with higher wages for workers in a marginalized mountainous region (Larson 2010; Bowen 2011). An additional instance is the Spanish Cabrales cheese whose DO production has enriched the local economy and has been a factor in a renewal of the region's economy through ecotourism (Rodríguez 2002). Similarly, in Italy, labor-intensive production of Parmigiano Reggiano (Parmesan) cheese contributed to the creation of more employment opportunities than industrial production of the same type of cheeses (Bowen 2015).

However, the benefits are not just economic. At the social level, the process of institutionally recognizing a DO requires collective action and mobilization on the part of producers to appropriate cultural heritage and increase self-esteem and motivations (Pomeón et al. 2011). The success of regulatory councils responsible for setting and implementing the regulations and standards that define DOs requires actors comprising the different parts of the agri-food chain to coordinate and cooperate (Sanz and Macías 2005). At the cultural level, given that a terroir is a combination of natural, physical conditions and the historical know-how of pertinent populations, codifying this knowledge and embedding it in local producer organizations allows for the maintenance of traditional production methods. Thus, DOs contribute to shoring up cultural heritage in the face of the dominant trend of homogenizing the global food industry (Renard 2010; Introduction, this volume). In some cases, they may also contribute to the diversity of rural landscapes and the conservation of natural resources.

In essence, the strategies generated by DOs belong to a set of market niches centered on different understandings of quality. Ultimately based on consumer preferences and valuing local production, they take entire sectors – such as wine production in Europe – away from the dominant industrial logic. Based on this local production, DO products enter the global market under more favorable conditions (Renard 1999). In other words, to a certain extent, they contribute to the plurality of the agri-food sector. Other case studies, however, have questioned the suitability of these strategies and their efficiency in achieving objectives such as integrated rural development and poverty alleviation (Besky 2014). The legal protection afforded to certain products does not uniformly or systematically reward the myriad of stakeholders in the agri-food chain (see Niederle, Wilkinson and Mascarenhas, this volume). Moreover, the control of the monopoly rent is the subject of disputes that entail the use of power (Linck 2018). Indeed, some aspects of DO policies may even exclude small-scale and/or pay only lip service to the improvement of working conditions and wages.

Benefits for the local population and producers depend on the institutional framework that recognizes and supports Dos, and this in turn is reflected in the power relations that, in each case, govern their construction. In European countries, the DO model has local, national and Europe-wide institutional support. In Mexico, there is an enormous variety of genuine artisanal products closely tied to their territories and produced in disadvantaged rural regions. Additionally, there is a need to generate local development tools to contain rural migration and the emptying of the countryside. In this context, Mexico has an institutional framework and a way to generate standards that are viewed as "good tools that produce bad results" (Larson 2010) as they reinforce the power of industrial sector elites at the expense of small-scale producers (Gaytán 2018; Bowen 2011, 2015).

Posing the question of the relevance of GIs as a tool for local socioeconomic development (Sekine, this volume) and employing the case of the DO for Mezcal in Mexico, this chapter discusses the relationship between the creation of DOs and the structure of socio-economic and political power. Specifically, we show that the exclusion of small rural artisanal producers caused by DO is not the result of oversight or the absence of institutions. Rather, it is associated with the actions of large-scale and corporate producers and the institutions responsible for the regulation of the DO. This group of elites uses DO as an instrument to preempt horizontal competition by seeking to impede the participation of other actors. In the first section, we will frame the case with a brief analysis of how the regulation of DOs functions in Mexico. Following a discussion of how the Mezcal DO was designed, we will illustrate the effects of the exclusion from the DO on small-scale rural producers from Teozacoalco, an Oaxacan Mixtec community. The attempt by corporate producers, supported by public authorities, to prevent small-scale producers, who are already excluded from the DO from using the word "agave," and the ensuing global response, also sheds light on some of the ways that the resistance movement against this exclusion has sought to appropriate its tradition and heritage.

In our analysis, we used a local agri-food systems (LAFS) focus. It analyzes the relationship between identity-based food products and their territorial anchorage, the activation of collective resources for their valorization and the territorial governance that such processes presuppose. The LAFS approach takes into account the relationship between food identities and classification mechanisms (institutional frameworks, regulations), the coordination between actors and collective actions, the management of cultural and natural resources and the knowledge and skills in play (know-how) (Muchnick 2012). The unit of analysis for these elements is the territory, which allows the analysis to be extended to other activities, such as tourism. For the development of this study, fieldwork was carried out at various intervals between 2013 and 2017 in Oaxaca, the Mixtec region and Teozacoalco. Interviews were conducted with a range of stakeholders: twenty-five producers from Teozacoalco, nineteen of which are both agave producers and small distillers and ten from the Central Valley, six of which produce both agave and process Mezcal; representatives from the municipal authorities; officials from the Mezcal Regulatory Council; and an official from the Ministry of Agriculture.

DOs in Mexico: institutional gaps and power relations

Sixteen products have DO status in Mexico. They include five distilled beverages: tequila (Jalisco, Guanajuato, Michoacán, Nayarit and Tamaulipas), Mezcal (nine states), *Bacanora* (Sonora), *Sotol* (Chihuahua, Coahuila and Durango), *Charanda* (Michoacán); three processed products (Veracruz coffee, Chiapas coffee, Morelos rice); five fresh products [the Ataúlfo mango from the Soconusco region (Chiapas), chile habanero from Yucatán, Yahualica chile de árbol (Jalisco), Papantla vanilla (Puebla) and Grijalva cacao (Tabasco)]. In addition to these foods and spirits, two handicrafts (Puebla Talavera pottery and Olinalá lacquerware from Guerrero) and one fossilized resin (amber from Chiapas) received DO status. As this chapter will illustrate, many of these DOs have not been able to create conditions necessary for local socioeconomic development and the emancipation of marginal producers.

The Mexican Industrial Property Institute (IMPI) grants DOs and, as the name implies, is primarily responsible for the protection of patents and trademarks (Pomeón et al. 2011). According to the Intellectual Property Law, once a DO has been approved, it remains the property of the Mexican state, and the only obligation of the IMPI is to protect it against imitations. The IMPI works under the jurisdiction of the Ministry of the Economy, which is also in charge of industrial development policies. The raw material–producing agricultural sector falls under the administration of the Ministry of Agriculture, Livestock and Fisheries (SAGARPA). SAGARPA, however, is not tasked with the promotion of products with territorial identity or DO policies (Pérez 2018). This situation explains the bias that authorities in charge of DOs have in favor of industrial outfits and activities (Pérez 2018). Accordingly, the role of the territory is hardly taken into account by the IMPI. Similarly, the fundamental relationship among the quality of the product, its characteristics and the geographical environment, or terroir, are downplayed. Here, terroir is understood as the combination of natural and cultural factors and the producers' know-how (Avelino 2006). The standards used for tequila – the most well-known Mexican DO and alleged success story – do not protect the relationship between the product and the place nor the practices that make the product unique (Bowen 2015). Such standards are in fact solely technical (e.g., the alcohol content and percentage of sugar), promote the industrial homogenization of the product and are thus primarily relevant to the now transnational industrial corporate elite. Most of the producers of the raw material, agave, are local small-scale farmers, and their knowledge is de facto not taken into account, nor is there protection for the diversity of species and natural resources that may be overexploited in the case of the (successful) expansion of production (Bowen 2011, 2015).

Hollowing out the concept of terroir of its original meaning, these denominations generally encompass overly extensive regions, such as coffee from the entire state of Chiapas (twelve regions and eighty-three municipalities) or from all of Veracruz. These overly broad, overly general DOs do not allow local diversity to be showcased, because they tend to homogenize rather than differentiate (Renard 2010). Additionally, they do not have the same level of recognition by roasters and consumers as that given to certain coffees with a bona fide territorial identity such as Cuxtepeques, Jaltenango (Chiapas), Pluma Hidalgo (Oaxaca) and Ixhuatlán del Café (Veracruz), none of which are DOs (Renard 2012).

Territorial names are also being "generified" by allowing the use of the word "type" after the indication (Larson 2010). Numerous "Cotija-type" cheeses that have nothing to do with those from the region of Sierra de Jalmich, "Oaxaca-type" cheeses made in Chiapas, and even "Manchego-type" cheeses from La Mancha in Spain are all cases in point. Incidentally, the latter instance led to a standoff in the renewal of Mexico's free trade agreement with the European Union (El País 2018). The dominant influence of the dairy industry is often employed to explain the overly permissive regulatory framework toward these imitations of genuine artisanal cheeses. The protection of corporate interests has prompted the IMPI to deny the DO application submitted by artisan producers and academics from the historic region for Cotija cheese, arguing that Cotija is a generic name. Petitioners were eventually permitted to use only a collective trademark owned by the producers that offers a lower level of protection (Pomeón et al. 2011; Pomeón 2007).

While a DO for Cotija cheese that meets all the requirements concerning its long historic roots, genuine quality and codified artisanal methods has not been recognized, in 2012 DO status was granted for the Ataúlfo mango from Soconusco despite the fact that the mango is not a species typical to Mexico – but to India – and the Ataúlfo variety is the result of a relatively recent technological innovation. The producers of the Soconusco Ataúlfo mango are entrepreneurs with political power and connections with public officials at local, state and national levels. They applied for the DO within a logic of commercial protection against competition from other regions of the country where the same fruit is produced (Velázquez 2017; Torres, Morales and Velázquez 2017). In the same vein, the extension of the tequila DO to the state of Tamaulipas was the result of pressure from an ex-minister of agriculture and grandson of a Mexican president who had set up a tequila agave (*Weber Azul*) plantation in that state. In fact, the IMPI violated its own standards by including in the tequila DO Tamaulipas, which is on the border with the United States and thousands of miles away from the Tequila region (Bowen 2015). In general, in Mexico, power relations and the logic of commercial protection take greater priority in the recognition of DOs than inclusive regional development. In this way, heavy lobbying of state authorities and the potential for product exports are essential factors in obtaining or receiving access to a DO (Pomeón et al. 2011).

Accordingly, a persistent trait of Mexican DOs is their being co-opted by elites with the support of public institutions (Gaytán 2018). This situation explains not only their design, tailored to these groups' interests, but also the watering down of the technical standards and permissiveness toward violations associated with these standards. An example of this latter issue is the granting of a request of tequila distillers to be allowed to include 49 percent alcohol from other sugars in order to match growing demand from the United States (Bowen 2011, 2015). Although the DO stipulates that all stages of production be contained in the protected region, when raw materials are scarce, industrial producers have padded their supply with agave from other regions and even other states, such as the Oaxaca mezcal-producing region. This DO violation has gone unpunished (Domínguez Arista 2016). As we will discuss in the comparable case of mezcal, control of the DO by a group of wealthy entrepreneurs with the support of regulatory institutions has meant that most of the traditional artisanal producers of this beverage have been excluded.

Mezcal and its DO

Mezcal is a traditional Mexican agave distillate produced in many regions of the country, using a process carried out according to the traditions of each location. The word *mezcal* comes from the Nahuatl *mexcalli*, which means "cooked maguey" (Serra and Lazcano 2016). It is not the name of a region but rather that of a production process, a generic name for a drink made by decocting and distilling agave, also known as maguey. By way of analogy, referring to a generic mezcal is tantamount to speaking of wine in generic terms; just as there are innumerable wines whose differences are tied to their respective terroir, there are many mezcals: tequila is a type of mezcal, as is *Bacanora*, both of which obtained their own DO. Unlike tequila, which has become an industrial and homogeneous product for the most part, mezcal continues to be artisanal. It has specific local characteristics due to the varieties of agave used and the different production practices (decoction, fermentation) that modify its flavor. Mezcal produced in the agroecological region of Tamaulipas (in the north of Mexico) is neither the same as mezcal from the region of Michoacán (in the west of the country), nor mezcal from Guerrero or Oaxaca (in the south), and within each region, there are significant differences.

In 1994 and through the IMPI, the Mexican government granted mezcal a DO. Originally limited to five states, it included Oaxaca, Guerrero, Durango, San Luis Potosí and Zacatecas. Subsequently, municipalities from the states of Guanajuato, Tamaulipas and Michoacán were added in 2001, 2003, 2012, respectively, as well as from the state of Puebla in 2015. This situation matured despite the fact that mezcal is actually produced in in at least twenty-six of the thirty-two states of the country (Colunga 2016). The decision to initially limit protection to five states stems from the industry elite that, through the Mezcal National Chamber of Commerce, applied for a DO for only these five states. The Mezcal National Chamber of Commerce is an association consisting of entrepreneurs of mezcal that operates with the ultimate goal of obtaining greater political power (Gaytán 2018; Bowen 2015). More than 80 percent of mezcal production in the protected states comes from the Central Valley region of Oaxaca, known as the Mezcal region, where the largest distilleries and producers are located. Representatives from this region control the Mezcal Regulatory Council (CRM). Created in 1997, the CRM is the regulatory institution in charge of DO certification, where, however, the representation of small artisanal mezcal producers is almost nonexistent.

A number of Central Valley producers have grown to be large companies and "success stories," giving them a more prominent role in the mezcal industry and greater weight in the dynamics of the Mexican institutions responsible for mezcal policy. They challenged each of the successive extensions of the DO to other states (Vega and Pérez 2018). In this context, the latest episode of the "DO war" occurred in August 2018, when the IMPI extended the DO to municipalities in the states of Mexico, Aguascalientes and Morelos. Producers from Oaxaca, led by the governor of the state, marched in protest through the streets of the national capital all the way to the offices of IMPI. They demanded a withdrawal of this extension under the argument that these producers have no tradition (CRM 2018). Additionally, it appears that the previous inclusion of Michoacán, which took years to be accepted, is being called into question, and a reversal is being sought (Quadratin 2018).

The DO for mezcal led to Mexican Official Standard 070 (NOM-070), which describes the characteristics and specifications of the production, bottling and sale of mezcal. Inspired largely by the standard for tequila, this standard proposes an industrial model that favors technical efficiency and a standardized product over biological and cultural aspects. This narrow, technical definition of quality is unable to capture the relationship between production and specific areas. It does require mezcal to be locally bottled but only specifies that 80 percent of the alcohol be derived from agave sugars and allows it to be produced using five types of agave. This latter point is even broader than tequila standards, which call for the liquor to be made only from agave Weber Azul. The MDO (Mezcal Denomination of Origin) covers a culturally, morphologically and environmentally disparate territory that subsumes mezcals as diverse as the regions it encompasses without recognizing their differences.

Insofar as the DO "confers exclusive agave planting and harvesting rights for Mezcal production in these areas" (Official Gazette of the Federation of Mexico, DOF 2012, 1), the declaration of the DO effectively rendered all authentic mezcals outside the boundaries as clandestine. Mezcals produced outside of the MDO can use only the technical label "agave distillates," which precludes them from using of the word *mezcal*, which is the ancestral name of these beverages. The problem is that the DO area was designed based on political-administrative boundaries (states and municipalities) rather than agroecological and historical boundaries. This DO excluded not only numerous culturally and environmentally important production areas but also their respective traditional small-scale producers. These authentic artisanal producers are de facto being pushed underground. They are also denied the other advantages that being in a region protected by the MDO entails, such as access to funding for state-sponsored mezcal programs. The Central Valley region is effectively given exclusive access to the millions of pesos of governmental support for this product. Furthermore, promotion schemes such as the "Mezcal Route," the most important tourist trail for mezcal, take only this one region into account. Extending the MDO to other regions would thin the distribution of government resources for mezcal retailers and producers, which in part explains the vested interest in denying any extension of the protected area. One example of the marginalization built into the design of this DO is the Teozacoalco community in the Oaxacan Mixtec region.

San Pedro Teozacoalco: an example of exclusion

Teozacoalco is a town settled in 1321 in the region of the Oaxacan Mixtec, with a tradition of mezcal production that can be traced back more than four generations and a well-earned fame in the region. The Mixtec is an ancestral mezcal production region and the spirit has symbolic and ritualistic meaning for its inhabitants (Domínguez Arista 2016). However, it was left outside of the boundaries of the MDO. The municipality of Teozacoalco is situated just outside the administrative border of the MDO. In this municipality, the residents have historically lived off the production of mezcal. This production is largely local in nature and is clandestine since the spirits are not registered with the CRM and since it can be legally called mezcal. This exclusion creates a number of additional problems. First, local producers do not bottle their products, and because they do not have a registered trademark, they create no "added value." Second, because production is not registered, intermediaries (known as coyotes) put pressure on local producers to sell at discounted prices. Coyotes usually resell local mezcals at higher prices, making small local distillers the weakest financial component in the production chain. Third, the agaves planted locally come from other regions and productions, including from tequila plantations. This situation has left local producers largely without raw material. Finally, local producers are the victims of counterfeiting. In Oaxaca, the state capital, but also in other cities, it is possible to find mezcals with the name of Teozacoalco despite the fact that they are not produced locally. Even distillers from other certified regions who are familiar with the reputation of the municipality have used the appellation *mezcal de Teozacoalco* on their label despite the fact that their products do not originate from this community or communities that are legally allowed to use the mezcal label. Additionally, Teozacoalco is not a registered trademark. Accordingly, no legal claims are possible. Moreover, this type of legal procedure is incomprehensible for Mixtec peasants since it is completely outside the sphere of their everyday life.

Another significant problem is the lack of organization and unity among mezcal processors and agave producers, which limits their ability to manage natural resources and maintain a consistent supply of raw materials – in this case, maguey. There are two ways of gaining access to maguey in this municipality. The first is to plant the individual plots of land each resident holds. However, not all farmers who plant maguey make mezcal. Some make a living by selling their maguey to any buyer leading to a shortage of raw material for local mezcal. The second is harvesting wild maguey from common lands. This wild maguey is the most sought-after variety because of its reputed superior taste and curative properties. Because it grows on common land, it is accessible to all residents. However, because there is no organization to regulate access, wild maguey has been harvested indiscriminately and depleted to virtual extinction.

Another assault on artisan producers: NOM 186 and *komil.*

As if banning artisan producers outside the DO from using the name mezcal was not enough, there also were attempts to prevent them from using the word *agave* in descriptions of their products. In essence, the idea was for corporate and large-scale tequila and mezcal producers to appropriate and privatize the word *agave* by transforming it from a common and scientific term to a specific name (Colunga 2012; Hernández 2016). In November 2011, the Minister of the Economy and the IMPI sponsored an initiative designed to permit the use of the word *agave* only to describe distillates produced and registered in DO territories (PROY-NOM-186-SCFI-2011). The supporting argument contented that this action would protect consumers from adulterated spirits as unregistered, non-DO distillers could label their products only "Agavacea aguardiente" or "distilled Agavacea"; given that these terms are incomprehensible to many consumers, they relegate these products to the margins.

Stating the percentage of agave on the labels was also banned. This created problems to producers of 100 percent agave spirits but benefited the tequila industry that was required to use only 51 percent of agave. In the weeks that followed, a controversy developed in the media and the Federal Regulatory Improvement Commission between supporters and opponents of the initiative. In favor of the project were transnational tequila conglomerates and the regulatory councils for tequila, mezcal and *Bacanora*. They argued that this initiative defended the reputation of the spirits and promoted consumer safety. In essence, they positioned themselves as saviors of the industry and the market (Gaytán 2018). Artisan producers, farming organizations, academics, bar and restaurant owners and other social actors with links to mezcal culture stated their opposition in Mexico and more than twenty other countries, including the United States, Australia and Germany. They objected to the standard in the name of transparency, fair competition, artisan crafts, tradition and the interests of small-scale producers (Colunga 2012; Gaytán 2018). Ironically, this campaign defended the original objectives of DO programs, such as protecting cultural heritage.

In February 2012, the regulatory authorities withdrew the initiative, which represented a surprising victory for small artisan producers and their supporters. However, the industry did not concede defeat, and another initiative was launched in 2015 (PROY-NOM-199-SCFI 2015). It proposed to use the word *komil* on the labels of all mezcals produced outside the DO and to prohibit the use of the word *agave* on all these labels (Hernández and de Jesús 2016). *Komil* means "intoxicating drink" in Nahuatl, the Aztec language. The supporting argument was the same as in the past: protecting the consumer from counterfeit and adulterated mezcals that do not actually contain agave or contain it in a lesser amount than suggested by the label. Despite this argument, NOM-199 would apply not only to low or non-agave beverages but also to all of those that were not in the DO. This action would de facto put authentic mezcal on the same footing as adulterated drinks.

Promoted by the same earlier defenders of artisanal mezcal, a significant opposition movement emerged. It coalesced around the slogan "it's called Mezcal, and we're going to defend it" (El País 2016; Hernández and de Jesús 2016). Letters were circulated, and the academic and cultural sectors were mobilized. The editor of a specialized mezcal website stated that "if this standard is approved, it would be a wholesale gutting of the cultural identity of many regions where Mezcal is traditionally produced for self-consumption or minor sales. It's a factor of inclusion and celebration; without Mezcal, there's no party" (El País 2016, B1). The word *komil* is not well-known, nor does it express any sense of embeddedness for the local population, unlike the words *agave* and *mezcal*, which are widely recognized. Because of this mobilization, a modified standard was adopted. The word *komil* was not used, and mezcals not included in the DO had to be called *agave aguardiente*.

Furthermore, the DO norm, NOM-070, was amended. Its final version established three types of Mezcals: mezcal, artisanal mezcal and ancestral mezcal. These categories do not provide a different protection from the DO; they are merely part of a marketing strategy. The first category, mezcal, refers to industrial mezcal. The second and third categories refer to ancient practices still used in the production of the distillate. Dividing mezcal in this way allows for the creation of market niches that give importance to words with symbolic value such as *artisanal* and even *ancestral*. These last two types of mezcal are highly valued by consumers and those that reach the highest market prices.

Conclusions

This case shows how the creation of the definitions of the GI are representations of political power. Supported by government institutions and opposing a broadening of the DO region, corporate producers of the Central Valley of Oaxaca excluded farmers and distillers from other mezcal-producing regions from using the DO label. The former deprived the latter of the use of the historical appellation *mezcal*, attempted to disallow the use of the word *agave* and finally forced them to use an unfamiliar name: *agave aguardiente*. As with most DOs in Mexico, this is an instance of politically powerful groups using a market tool to safeguard their interests. Ultimately, these actions led to the preservation of the monopolistic position of corporate producers.

The design of the mezcal DO is paradoxical. The area of the appellation is too narrow:, because it excludes many genuine producers. Simultaneously, it is too broad, because it erases local differences tied to biological and cultural diversity (Bowen 2015). One possible solution, comparable to the situation for spirits in Europe, would be to leave the name mezcal as a generic name and establish sub-denominations tied to the different territories, just as is the case of tequila and *Bacanora*. This solution would allow for the inclusion of currently excluded traditional producers and provide a separate opportunity for the creation of more exclusive DOs for well-known regions such as Teozacoalco.

Another alternative would be to restructure DO regulation and steer it toward the socioeconomic development of rural areas and marginalized producers. Besides the challenges of developing a suitable policy to accomplish this objective, the prospects for such an endeavor seem low, if not virtually impossible, given the neoliberal approach employed by Mexican administrations over the last thirty years. Affected by the influence of the United States on Mexican economic affairs, this government policy has favored foreign direct investment over the integrated development of marginal regions (Bowen 2011; Pomeón et al. 2011). The alternative implementation of a European-style DO policy is thus quite improbable given Mexico's membership in the North American Free Trade Agreement that entails a posture that does not support a sui generis DO model.

The relative victory achieved by the defenders of traditional mezcal against NOM 186 and NOM 199 was made possible thanks to the international relevance of this beverage in the cultural media and among aficionados in Mexico and other countries. It was also due to the active support from stakeholders positioned downstream in the agri-food chain (bars and restaurants). This support, while effective in this case, is probably insufficient to raise the profile for many other traditional Mexican products, such as real Cotija cheese, which did not obtain a DO despite efforts from a broad community of academics. Furthermore, this loose support movement is an indicator that there are no dedicated organizations defending territorially identified traditional products in Mexico. Accordingly, it appears unlikely that a stronger movement could emerge in the short term in a country where large segments of the population struggle to satisfy their basic need for food. In essence, facing a hostile institutional framework and without social pressure from organized, informed consumers groups and democratically organized producers interested in preserving local foods, it seems that putting DOs to work for small-scale producers will be difficult. This case relativizes the efficiency of strategies based on the qualification of products and consumer preferences as, when there is an uneven distribution of power among the stakeholders, weaker segments are either silenced or co-opted by the market-dominant actors. Despite this, resistance by vested producers from Oaxaca and the CRM to the extension of the mezcal DO to other states that arose in August 2018 by, shows that the DO of mezcal remains a contested terrain, entailing local mobilization and the continuing search for political support.

Bibliography

Avelino, Jacques. 2006. "Denominaciones de Origen e Indicaciones Geográficas: Fundamentos y Metodologías con Ejemplos de Costa Rica." In *El Cafetal del Futuro*, edited by Jürgen Pohlan, Lorena Soto Pinto and Juan Francisco Barrera, 119–40. Maastricht: Shaker Verlag.

Besky, Sarah. 2014. "The Labor of Terroir and the Terroir of Labor: Geographical Indications and Darjeeling Tea Plantations." *Agriculture and Human Values* 31 (1): 83–96.

Bowen, Sarah. 2015. *Divided Spirits. Tequila, Mezcal, and the Politics of Production*. Berkley: University of California Press, EUA.

———. 2011. "Las Indicaciones Geográficas como Estrategia de Desarrollo Territorial en el Sur Global: el Caso del Tequila." In *Los Sistemas Agroalimentarios Localizados en México*, edited by Gerardo Torres Salcido, Héctor Alejandro Ramos Chávez and Mario del Roble Pensado Leglise, 59–84. Mexico: CIICH, UNAM.

Colunga-García Marín, P. 2016. "Se llaman mezcales y están hechos con Agave, no engañen al consumidor." *La Jornada*, Saturday April 16, p. 11.

———. 2012. "La Desaparición de los Mezcales Artesanales Tradicionales." *La Jornada*, Saturday April 21, p. 12.

CRM. 2018. *Consejo Regulador del Mezcal*. Mexico.

Domínguez Arista, Rodolfo. 2016. "Los Límites de la Denominación de Origen y la Producción Artesanal del Mezcal: El Caso de Teozacoalco, Oaxaca." MA Thesis in Rural Sociology, Universidad Autónoma Chapingo, Mexico City.

El País. 2018. "México y la UE, una Guerra por el Queso Manchego." Online Document. Accessed January 10, 2018. https://elpais.com/economia/2018/01/08/actualidad/1515434055_646189.html.

———. 2016. "'Komil', la Palabra que los Amantes del Mezcal no Quieren Oír." Online Document. Accessed February 25, 2016. https://verne.elpais.com/verne/2016/02/25/mexico/1456362020_194480.html.

Gaytán, Marie Sarita. 2018. "The Perils of Protection and the Promise of Authenticity: Tequila, Mezcal and the Case of NOM 186." *Journal of Rural Studies* 58: 103–11.

Larson Guerra, Jorge. 2010. "Indicaciones Geográficas, Competencia Leal y Desarrollo Rural." *La Jornada del Campo*, February 13, p. 9.

Linck, Thierry. 2018. ¿Constituyen las Denominaciones de Origen Palancas para un Desarrollo Territorial Sustentable y Justo? In *Saberes de Origen. Experiencias de México y Centroamérica*, edited by Paplo Pérez Akaki, Alma Amalia González and Wilson Picado, 27–40. Mexico City: UNAM.

———. 1999. "Prefacio." In *Los Intersticios de la Globalización: un Label (Max Havelaar) para los Pequeños Productores de Café*, edited by Marie-Christine Renard, 15–29. Mexico City: CEMCA.

López, Hernández, and José de Jesús. 2016. "Carta Abierta al Maestro Mario Emilio Gutiérrez Caballero, Titular de la Comisión Federal de Mejora Regulatoria." Online Document. Accessed January 18, 2018. www.academia.edu/22299844/Oposici%C3%B3n_a_la_NOM_199_KOMIL.

Muchnick, José. 2012. "Sistemas Agroalimentarios Localizado: Desarrollo Conceptual y Diversidad de Situaciones." In *Sistemas Agroalimentarios Localizados. Identidad Territorial, Construcción de Capital Social e Instituciones*, edited by Gerardo Torres Salcido and Rosa María Larroa, 20–35. Mexico City: CIICH-FCPS-UNAM.

Official Gazette of the Federation of Mexico, DOF. 2012. *Diario Oficial de la Federación*. Online Document. Accessed January 10, 2018. http://dof.gob.mx/nota_detalle.php?codigo=5278677&fecha=22/11/2012.

Pérez Akaki, Pablo. 2018. "Denominaciones de Origen e Instituciones: Análisis de Las Experiencias Mexicanas." In *Saberes de Origen: Experiencias de México y Centroamérica*, edited by Paplo Pérez Akaki, Alma Amalia González and Wilson Picado, 41–68. Mexico City: UNAM.

Pomeón, Thomas. 2007. *El Queso Cotija, México: Un Producto con Marca Colectiva Queso Cotija Región de Origen. Proceso de Adquisición de una Denominación de Origen*. Mexico City: FAO-IICA.

Pomeón Thomas, Esteban Barragán, François Boucher, and Fernando Cervantes. 2011. "¿Denominación de Origen o Denominación Genérica? El Caso del Queso Cotija." In *De la Leche al Queso: Queserías Rurales en América Latina*, edited by François Boucher Boucher and Virginie Brun, 47–71. Mexico City: IICA, CIRAD, Miguel Ángel Porrúa.

PROY-NOM-199-SCFI. 2015. "Opposition to Nom 199 for Agave Distillates." Online Document. Accessed March 9, 2015. www.tequilainterchangeproject.org/2015/03/09/nom199/.

Quadratin, Agencia. 2018. "Buscan Expulsar a Michoacán de Mezcaleros con Denominación de Origen." Online Document. Accessed April 15, 2018. www.quadratin.com.mx/principal/buscan-excluir-a-michoacan-de-mezcaleros-con-denominacion-de-origen/.

Renard, Marie-Christine. 2012. "Los Retos de la Valorización de un Café de "Origen": el Caso de la Reserva del Triunfo, Chiapas." In *Sistemas agroalimentarios localizados. Identidad territorial, construcción de capital social e instituciones*, edited by Gerardo Torres Salcido and Rosa María Larroa, 117–33. Mexico City: CIICH-FCPS-UNAM, Juan Pablos.

———. 2010. "Proteger los Alimentos y Bebidas Tradicionales." *La Jornada del Campo*, February 15, p. 11.

———. 1999. *Los Intersticios de la Globalización: un Label (Max Havelaar) para los Pequeños Productores de Café*. Mexico City: CEMCA.

Rodríguez Gómez, Guadalupe. 2002. *La Denominación de Origen y el Mercado de la Distinción*. Mexico City: CIESAS-SAGARPA.

Sanz Cañada, Javier, and Alfredo Macías Vázquez. 2005. "Quality Certification, Institutions and Innovation in Local Agri-food Systems: Protected Designations of Origin in Olive oil in Spain." *Journal of Rural Studies* 21: 475–86.

Serra Puche, Mari Carme, and Jesús Carlos Lazcano Arce. 2016. *El Mezcal, una Bebida Prehispánica: Estudios Etnoarqueológicos*. Mexico City: UNAM.

Torres Salcido Gerardo, David Aarón Morales Córdova, and Víctor Manuel Velázquez Durán. 2017. "Gobernanza Territorial e Innovación Social en las Áreas Rurales. Dos Estudios de caso." In *México Rural Ante los Retos del Siglo XXI, Tomo IV Mercados Agrícolas Globalizados*, edited by Armando Sánchez Albarrán and Estela Martínez Borrego, 97–116. Mexico City: AMER.

Vega Vera, Nadia Viridiana, and Pablo Pérez Akaki. 2018. "Denominación de Origen Mezcal en Oaxaca. Bebida Artesanal, Tradición y Cultura." In *Saberes de Origen. Experiencias de México y Centroamérica*, edited by Paplo Pérez Akaki, Alma Amalia González and Wilson Picado, 253–78. Mexico City: UNAM.

Velázquez Durán Víctor Manuel. 2017. "Conferencia Sobre Gobernanza de la DO del Mango Ataúlfo en la Región del Soconusco." Paper Presented on June 2, 2017 at the Conference on "Gobernanza y Alimntos." Organized by CIALC, UNAM. Mexico City, Mexico.

WIPO. 1958. "Arreglo de Lisboa." Online Document. Accessed April 2018. www.wipo.int/lisbon/es/general/.

11 Whose labor counts as craft? Terroir and farm workers in North American craft cider

Anelyse M. Weiler

As our tour bus full of craft cider enthusiasts and farmers winds its way past the snowy, rolling acres of semi-dwarf apple trees, our guide picks up the microphone. She stresses that the cidery we're about to visit isn't really set up yet as a public tasting room. We would do well to use the restroom on the bus. As we're approaching the main farm building, I notice two rudimentary-looking mobile homes with a black pickup truck parked out front. I'm immediately reminded of visiting the temporary homes of Latinx farm workers in Canada as part of a grassroots organization that provides community support and labor rights advocacy. I decide, however, that I'm jumping to conclusions. Surely, Latinx farm workers wouldn't be living year-round in this isolated rural area, and orchard labor demands would presumably be low in the middle of winter.

Inside a chilly, high-ceilinged garage, a white farmer in his thirties with a beard and a trucker hat describes his family farm's two-hundred-year history and recent evolution toward cider. As he speaks, a younger female staff member pours us cider samples from beneath a small pop-up tent, setting the glasses onto a blue gingham tablecloth. The makeshift bar creates the feel of a farmers market within an office space, fermentation facility and storage shed that is clearly not designed for tourists. Although the temperature has dipped below freezing, the farmer invites anyone who is interested to join him on a tour of the cider orchard. Some twenty of us follow the farmer, shuffling past empty bottles of diesel oil, out the back door of the garage and beyond a rusty tin shed with signs warning, Peligro: Pesticidas. No entre (Danger: Pesticides. Keep out).

As we're finishing up the tour of the leafless apple trees, I squint ahead through snowflakes that have begun to lazily tumble over the orchard. One of the farmers seems to have struck a jaunty pose up in an apple tree, as though to entertain our group. When I get closer, however, I notice that it's actually two farm workers bundled up in fleece jackets and pruning the trees. A tour attendee records one of the men on her smartphone.

Instinctively, I worry to myself, if the worker is undocumented, this video could expose him to the border patrol. Earlier, I had summoned the most casual tone of voice I could muster to ask how orchardists in his region manage challenges with sourcing labor. He responded that in his region, the government-administered H-2A temporary agricultural worker program had not worked well. When US agricultural employers report labor shortages, they may use an H-2A visa to hire non-immigrant workers from outside of the country on a seasonal or temporary basis. Although the farmer didn't spell it out, this may mean that farmers instead hire workers without papers.

After the tour, the attendee posts the pruning video on Instagram. She tags the farm location and emphasizes her admiration for "the skill and grace of it all." The farmer who had toured us around the orchard weighs in on her comment thread, declaring, "There's no apple, no cider without these guys!" "For the record, nobody was making them prune in the snow. They're paid by the tree, and that's why they were out there." He notes that his farm provides health benefits to workers, explaining that someone on the tour had been passing judgment about farm worker rights. It is unclear whether he was referring to me; beyond inquiring how farms in the area address the challenge of hiring workers, I had not articulated any criticisms or reproachful questions. He concludes by apologizing for sounding defensive. "Thanks for giving their hard work the recognition it deserves!"

Introduction

In this chapter, I consider how the fledgling North American craft cider industry has sought to elevate the value of a new niche processed agri-food product by telling a story about how cider is intertwined with rural livelihoods and local landscapes. Given that North America's craft cider industry is in the early stages of development and at an even earlier stage of pursuing the formalized recognition of terroir, understanding how the industry accounts for human labor within terroir helps to illuminate the extent to which GI might become emancipatory for all those engaged in the cider production chain. The popular concept of terroir is based on the premise that the taste characteristics of an artisanal food or drink are determined by the product's unique environmental context, along with the farming practices of humans who cultivate it (Cappeliez 2017). More broadly, place-based foods and associated labeling systems like GI are premised on the idea that a price premium can protect certain agricultural practices and values from globalized market pressures such as consolidation and standardization (Bowen 2015). Agricultural tours like the one I described at the outset of the chapter are a powerful venue for instilling a sense of personal connection to a product, a place and the people who made it. Visiting a rural cidery is premised on offering consumers a transparent view of how their food or drink was made and the opportunity to verify firsthand whether product claims about quality, ecological sustainability or other business practices appear accurate.

What struck me about this encounter is that it offered a rare glimpse beyond the performative curtain of place-based agritourism in the United States. Unmediated interactions with Latinx farm workers are off script from the standard meet-your-white-farmer experience. In the case of a fair-trade GI product like Darjeeling tea, tea pickers' situatedness within the plantation landscape forms a prominent part of product branding and tourism. Besky (2014) describes how actors who stand to benefit from the success of Darjeeling tea fetishize female plantation workers' "delicate" manual labor and view it through a nostalgic filter that romanticizes a painful capitalist and colonial legacy. She builds on Guthman's (2014) concept of a culturally conservative agrarian imaginary that imbues US organic agriculture with a sheen of virtuousness while obscuring issues such as the industry's dependence on racialized, low-wage workers with weak bargaining power. Similarly, Sekine (this volume) highlights how the viability of GI Nishio Matcha in Japan hinges on local, mostly elderly female harvest workers, whose wages are extremely low relative to service sector or manufacturing workers.

In the growing market for North American craft cider and agritourism, the labor of Latinx orchard workers is not part of the selling point. For instance, it would be considered strange for cider enthusiasts to debate whether the tree-pruning technique of mostly non-local orchard workers affects the taste of craft cider from a given farming locale. Amid rising xenophobic and nativist sentiments, the topic of immigrant or migrant farm workers has become increasingly polarizing. A highly unequal employer-employee dynamic does not lend itself straightforwardly to romanticized product marketing. On the cidery tour, racialized orchard workers' presence in the agrarian landscape became a problem to be normalized to consumers, like the lack of public restrooms, or having to host an artisanal cider bar in an unglamorous garage. The farmer's expression of gratitude for his employees' "hard work" could be read as an attempt to smooth over the obvious hierarchies of race, class and citizenship at hand. Gratitude, however, cannot be taken to the bank.

The political-economic context of farm labor migration within the apple and cider industry presents considerable challenges to realizing emancipatory outcomes through GI. While labor protections for farm workers are already weaker than those in many other industries, immigration policies affecting undocumented and temporary migrant workers further undermine their access to human and labor rights (Gray 2014; McLaughlin, Hennebry and Haines 2014). Market liberalization processes such as the North American Free Trade Agreement implemented in 1994 have had uneven effects on the viability of agrarian livelihoods within migrant-sending countries such as Mexico (Eakin et al. 2018). Many scholars argue that free trade agreements and structural adjustment policies have played a significant role in increasing temporary farm labor migration by undermining local livelihoods in workers' home countries (Otero 2011; Minkoff-Zern 2014). Proponents of GI argue that it can ward off the harmful effects of neoliberal capitalism on small and medium-size producers, ecosystems and traditional food cultures (see Bonanno, this volume). As a market-based solution, using GI to increase farmers' competitiveness through a marker of product distinction is compatible with strategies to remain competitive by hiring low-wage migrant workers. It remains an open question whether market-based solutions can align with the political-economic change required to equalize entrenched asymmetries of power and produce emancipatory outcomes for farmers and farm workers alike.

A manual-labor renaissance

Craft cidermaking falls into step with what Ocejo (2017) describes as an occupational renaissance in the new economy, in which savvy young people are recasting traditionally low-status manual work into trendy, specialized and personally meaningful careers. In contrast to dominant trends toward standardization and automation, craft occupations allow people to bring to bear their highly specialized knowledge and create small volumes of a unique, tangible good. Jones (2016) contends that contemporary artisans such as cidermakers, vintners and butchers are driven by the moral idea that crafting "good" material products simultaneously constitutes them as "good" producers who are bringing about a better world. As a value-added farm product, cider offers an enticement for younger producers and a potential lifeline for older orchardists who struggle with a volatile commodity market. Many hope regional craft cider will help save family farms and enable viable long-term livelihoods for young craft producers. It is unclear, however, whether cider can tread beyond the waves of fickle market trends and fulfill the hopes people have pinned on it. Moreover, existing research on the clustering of craft livelihoods in the United States among highly educated and predominantly white men (e.g., Ocejo 2017) raises questions about whose manual labor is celebrated as part of this occupational renaissance and how the material benefits are dispersed.

Food studies researchers have illuminated how food labor – from the field, to the cash register to the kitchen sink – is devalued and made invisible across lines of gender, race and class (Cairns and Johnston 2015; Coplen 2018). For premium value-added foods like craft cider, many raw ingredients are produced through automation or farm jobs that are divided into narrow, repetitive manual tasks. Farm workers in affluent countries are often racialized and from the Global South. They may be hired through government-administered labor-migration programs, undocumented streams or domestic permanent resident populations. Whereas seasonal migrant farm workers in Canada are commonly hired through a government-administered labor-migration program, just under 50 percent of US crop farm workers are undocumented (USDOL 2016, i). In both cases, assigning workers a precarious immigration status gives the state and employers tremendous power over workers, who may be repatriated if they resist (Bonanno and Cavalcanti 2014; Weiler, McLaughlin and Cole 2017).

Farm work frequently involves weak access to social protections, exemptions from numerous labor standards, low social prestige and barriers to workplace democracy (e.g., unionization) (Weiler, Otero, and Wittman 2016; Paz Ramirez and Chun 2016). The 2013–2014 U.S. National Agricultural Workers Survey reports that farm workers had a mean and median income from farm employment within the range of US$15,000 to US$17,499 and that 30 percent of farm workers had a total family income below the poverty level (USDOL 2016, iii). In Canada, the employment income for agriculture and horticulture workers is C$15,286 (median) and C$21,700 (average), compared to C$36,693 (median) and C$48,930 (average) for all occupations (StatCan 2017a). Along with workplace hazards such as agrochemical exposure, researchers have documented the prevalence of a lack of on-farm sanitary facilities and overcrowded, substandard farm worker housing (Hennebry, Preibisch and McLaughlin 2010; Arcury et al. 2010). In short, farm work is often constructed as a "bad" job.

As a strategy to promote "good jobs" and socioeconomic development for people from the Majority World who have been marginalized by contemporary expressions of capitalism, GI labeling may offer highly uneven benefits and can exacerbate preexisting inequities. For instance, when the Mexican government introduced a type of GI, denomination of origin (DO), that defined the territory within which mezcal could be produced, many traditional producers in parts of Mexico outside of that territory were no longer legally permitted to make mezcal (Bowen 2015; Renard and Arista, this volume). In the United States, GIs are treated as a subgroup of trademarks governed by the U.S. Patent and Trademark Office. Labeling requirements for alcohol fall under the authority of the Alcohol and Tobacco Tax and Trade Bureau. Despite the emphasis on private law, Le Goffic and Zappalaglio (2017) point out the role of publicly funded US institutions in protecting GIs. In Canada, the Trademarks Act allows GI protection for agricultural and food products through "protected GI" for agricultural products and food (i.e., international protection via trade agreements), official marks (protection in the domestic market) and certification marks (market protection only in the countries where an organization has requested to register the mark) (Watkin 2018). GIs fall under both federal and provincial jurisdiction. The province of Quebec, for instance, adopted an Act on Reserved Designations and Added Value Claims in 2006 that bears similarity to certain legislative approaches to GI in Europe. Quebec's law created designation categories for (1) protected geographical indication (PGI); (2) protected designation of origin (PDO); 3) designation of specificity; and 4) production methods, such as the organic designation (Ben Hassen and Tremblay 2016). Watkin (2018) argues that the ostensible lack of harmonization between provincial and federal governments has created a challenging context for endogenous GIs to thrive.

Within the globalized cultural context of food and drink, Cappeliez (2017) has shown how wine actors have culturally translated the concept of terroir from French to Canadian wine regions. She argues that while the normative principles associated with terroir (e.g., terroir as a unique and romanticized place) tend to remain constant across cultural contexts, aspects of terroir that are tied to place (e.g., human relationships to nature and practices on the land) are locally contingent. Less well understood is how the benefits of place-based craft products flow to people who cross borders from poorer to wealthier countries. Amid efforts to link craft cider to specific places and construct artisanal livelihoods as prestigious, how does the craft industry account for its dependence on workers who are not from those places and are employed in so-called bad jobs? How do agrarian imaginaries play out in a local artisanal industry that is deeply dependent on non-local migrant or immigrant workers? On both a symbolic and material basis, how does the renaissance of place-based, processed artisanal foods and manual occupations reward those engaged in producing raw ingredients?

This chapter draws on preliminary insights from ongoing research on craft cider in Canada and the United States. My key methods have entailed qualitative, in-depth, semi-structured interviews and participant observation. Participants throughout British Columbia (BC), Canada, have included thirty English-, French- and Spanish-speaking farm workers from a wide range of class backgrounds, thirteen cider producers and six non-cider-producing farmers, as well as fifty-two shorter interviews with consumers at apple festivals. I also draw on an interview with a cider producer in Washington state. In addition, I have engaged in participant observation as a hired worker on a small apple orchard in British Columbia, as a participant in farm tours and cider education workshops in the United States, and as an attendee at a national US cider conference. Throughout my research, I also took fieldnotes on my frontline volunteer work with farm worker justice organizations in British Columbia and Ontario, Canada. Activities included everyday support such as translating during medical appointments, helping workers exit abusive workplaces and engaging in a solidarity visit with a farm worker encampment in Washington state.

After sketching the historical and contemporary context of North American craft cider production, I will present data on how the craft cider industry makes sense of terroir and how hired workers fit into this preliminary picture. I conclude by evaluating the extent to which place-based labels such as GI might promote socioeconomic development for farmers and farm workers in the craft cidery industry.

Cider's history in North America

In the United States, most early settlers grew apple trees from seed (Futrell 2017). Because apples grown from seed are often too bitter to eat, they filled colonists' cups to the brim with hard cider. European colonists had brought to North America both apple seeds and their fondness for this form of alcohol, which people believed was healthier than untreated drinking water and even medicinal (Aaron and Musto 1981). Historians have also documented the carryover of European traditions of cider apple cultivation and fermentation in the early to mid 1700s in what is now known as Canada, particularly in what was then called New France (Fortin 2000; Jordan 2016). With the temperance movement in North America, alcohol and drunkenness became perceived as a threat to social order (Aaron and Musto 1981, 136). Kerrigan (2012) argues, however, that the decline of North American hard cider was chiefly because of economics. The market favored commercialized orchards that grew sweet apples ready to eat raw rather than tannic cider apples.

Alongside the waning of US cider, labor dynamics on apple orchards were also shifting. In the early 1700s, US family homestead orchards were largely self-sufficient (Futrell 2017). Larger American plantation orchards depended on sharecroppers and enslaved people (e.g., USDA 1860). This history has shaped subsequent racial and class dynamics of US apple production writ large, along with regional immigration dynamics. For instance, in the late 20th century, farms in the major apple-producing state of New York shifted from hiring mainly black workers – including both African American and Caribbean guest workers – to hiring Latino farm workers (Gray 2014, 102). Racialized people of Latin American origin are disproportionately represented among US farm workers today (USDOL 2016, i).

Craft cider today

In the contemporary settler-colonial context, apple growers in Canada and the United States are integrated into a globalized apple commodity market, and producers of all sizes face a dizzying array of competitive pressures. These include pressures to consolidate businesses and farmland into fewer hands and vertically integrate. Retailers and wholesalers hold immense power to impose private standards and shape prices, while institutions that own intellectual property rights such as trademarked "club" apple varieties like SweeTango can compel all producers in a region to abide by their restrictive rules (Futrell 2017). Amid the "cost-price squeeze" and other dynamic pressures, farmers often report labor shortages, asserting that it would be cost-prohibitive for their businesses to attract local workers by improving wages and working conditions (Weiler, McLaughlin and Cole 2017). Simultaneously, labor advocates, eaters and even some farmers have raised concerns about human and labor rights violations associated with both government-administered farm labor-migration programs and undocumented migration (Binford 2013; Futrell 2017). The confluence of these dynamics makes the prospect of growing apples for the commodity market broadly unappealing to either children of growers or new entrants from non-farming backgrounds. The number of Canadian farms is dwindling, with larger and more capital-intensive farms making up those that prevail (StatCan 2017b). In 2016, the average age of farmers in Canada was fifty-five. Similarly, the average age of US farm operators ranged from fifty-five to fifty-six between 2007 and 2012, which is consistent with a thirty-year trend of aging farmers (USDA 2014).

Amid the considerable challenges of maintaining a viable apple orchard today, craft cider has been pitched as a lifeline to save family-owned orchards. This parallels a broader pattern in which farms are attempting to stay afloat through value-added farm products, agritourism and off-farm income. Unlike the mainstream commodity apple market, hard cider allows farmers to add value to raw farm goods through processing and product differentiation. However, capitalizing on a food and drink trend can still subject farmers to the whims of the market, fickle consumer preferences and pressures from large-scale agri-food companies.

Corporate "craft" cider and boundary making through the taste of place

Paralleling the trend in craft beer, the market for hard cider exploded in the latter part of the first decade of the 2000s. After growing from US$43.5 million in US off-premise sales outlets in 2008 to a peak of US$535.7 million in

2015, this trend appears to have diminished slightly or hit a plateau; cider sales were US$469.8 million in 2017 (fieldnotes from Nielsen presentation 2018). Large beer companies such as MillerCoors, Heineken and The Boston Beer Company have capitalized on the trend. Some large companies have straightforwardly positioned some of their ciders as inexpensive and generic alcohol options, pejoratively known as alco-pops. Simultaneously, large beer companies have introduced product lines marketed as craft and artisanal. Beyond appealing to aesthetic and discursive repertoires that are typical of smaller-scale craft cideries, large cider-producing companies have mimicked processing techniques that signal product uniqueness and authenticity, such as fermenting in oak barrels and with wild yeasts. Some have also acquired smaller cideries or developed collaborative product lines with cideries that are perceived as genuinely artisanal and traditional. For instance, the largest US cider brand, Angry Orchard, developed a collaborative cider style in 2018 with Oliver's Cider, a highly regarded English craft cidery. Cider brands from large companies make up an estimated 75 percent of US cider sales. Still, premium regional and local cider brands have grown from 8.8 percent of cider sales in 2014 to 25.2 percent in 2017 and appear to be driving growth in the cider market (fieldnotes from Nielsen presentation 2018).

Small and medium-scale craft cider interviewees in my research project tended to express ambivalence about ciders from larger-scale producers. Some interviewees felt that so-called alco-pop ciders could serve as a "gateway" beverage that encourages consumers to subsequently explore niche ciders from small- and medium-scale producers. At a national US cider conference, a speaker presenting the latest cider market research expressed that growth in the cider market by large beer companies creates consumer "category awareness" and contributes to product innovation and said that "a rising tide lifts all boats" (fieldnotes from Nielsen presentation 2018). Nonetheless, other cidermaker interviewees resented what they felt was a form of co-optation by producers who appropriated the marketing language of craft cider but did not exemplify qualities that they felt typified craft cider. They felt genuine craft cider included features such as small- to medium-scale production, selling within short supply chains, partnerships with independent local upstream and downstream markets and forswearing production techniques such as adding refined sugar to increase alcohol content or using apple juice concentrate.

When large-scale companies muddy the waters of what counts as "craft" cider, this presents small- and medium-scale farm-based cideries with the risk of losing a marker of product distinction. Simultaneously, the growing market for premium regional ciders presents an opportunity for defensive boundary making by foregrounding the unique people and place behind a craft cider. Agritourism can further solidify consumers' association of a product taste with specific people and place. In British Columbia, a cidery that is situated on an orchard and grows some of its own apples receives a significantly favorable tax rate compared to a cidery that is classified as "commercial" (e.g., an urban cidery that sources apples or juice from external suppliers)

(BCLDB 2018). Consequently, many BC cideries draw on agritourism and place-based marketing.

Some craft cider producers have formalized the boundary-work of marketing place and taste by pursuing various forms of certified labeling, including GI. One of my interviewees explained how the Northwest Cider Association, a trade organization of producers in the US and Canadian Pacific Northwest, had been pursuing a regional seal to distinguish their members' products based on the local sourcing of ingredients. He indicated that amid producers' conflicting interests, however, this discussion had stalled because of a lack of consensus over the detailed content of the seal (interview with Washington state cider producer, 2018). In the case of GI, labeling for cider has been almost exclusively limited to Europe, specifically in France, Spain, the United Kingdom, Germany and Finland (Fabien-Ouellet and Conner 2018). An exception is among ice cider producers in Quebec. Quebecois producers lobbied the provincial government for over a decade to realize an exclusive, legally recognized label that protects producers' know-how (savoir faire) and is linked to a specific territory (fieldnotes 2018). In 2014, the Quebec Ministry of Agriculture formalized a PGI status for Quebec ice cider (CARTV 2014). Cider producers in US states such as Vermont have also indicated a strong interest in developing a GI label to heighten consumers' ability to differentiate among hard ciders and increase product demand (Fabien-Ouellet and Conner 2018).

A producer with a BC cider company that launched its first product in 2015 expressed the following regarding terroir designation for that province:

> You talk about terroir in the wine world and I think that's totally a relevant conversation for cider as well. . . . But I guess what I want to see is people picking up and having a glass of cider that's from BC and being able to talk about what that means, being able to say, "Hey, this tastes like it's a BC style."
>
> I think we need to be able to define more clearly what BC cider, BC craft cider, is. I think we need to define that as an industry. And then we need to create clarity on what that is outside of our industry, to our consumers. So, place of origin is huge. There's no VQA [Vintners Quality Alliance] equivalent in cider and I think that is a model to possibly look at. I think that eventually we'll be, should be talking about an appellation system not dissimilar to what is happening with BC wine or is in the works for BC wine and what has been in existence for a long time, in France for example.
>
> [. . .]
>
> Our industry's just so young, we're still figuring it out, and I think everyone is pretty much focused on cider making and trying to see their businesses succeed and grow. But I think as we grow as an industry, we'll have more time and more dedicated bandwidth to figure out some of those bigger conversations on what BC craft cider is and working with the government to help define that.
>
> – Interview with BC cider producer 2017

Formal GI designation is thus far limited, in part because the US and Canadian craft cider industry is still incipient. Many producers are focused on the early stages of developing their individual enterprises, retail partnerships and consumer bases. Still, some producers appear to be in the nascent phase of defining regional identities and conceptions of terroir that might underpin future GI efforts, should they even come to pass. For instance, tannic, cider-specific apple varieties that provide complex flavors can sometimes be difficult to access in the Pacific Northwest, and in any case, many producers felt local consumers had not yet learned to appreciate tannic ciders. To add flavor and place specificity, many ciders in this region are infused with unique, local, wild-harvested ingredients (e.g., native rosehips, salal berries and evergreen tree needles). It is common for products to be marketed with a general reference to local ecological landmarks such as proximity to the ocean or a specific body of freshwater, although a smaller number of enterprises emphasize the role of wild yeasts, soils or other natural features as defining their cider's geographically rooted taste. Moreover, some producers have formed regional associations to set quality standards, assist one another with farming and production skills and cooperate on product marketing, all of which would be a crucial precursor to formal GI designation.

Terroir and human labor

Many of the craft cider producers I interviewed felt that both the natural and human factors underlying cider production were significant in shaping the beverage's taste. Apple variety selection loomed especially large in terms of factors that producers felt were significant in shaping the taste of a place. In many cases, cidermakers were pragmatically drawing on whatever dessert apple varieties happened to be growing abundantly in their region due to factors such as market demand, the history of intellectual property dynamics (e.g., research and breeding rights) and suitability to local growing conditions. Some observed that cider apples from Europe tasted different in their region due to distinct growing conditions. One producer from interior British Columbia described differences in cider produced in his region even compared to the coastal area of the province due to distinct weather and soil:

> Yeah, I think there's definitely a terroir thing because I know when you look at ciders from the coast, they're usually going to be a lower alcohol content than ciders from up here just because of the weather. Yeah, I think it definitely makes a big difference being here.
>
> – Interview with BC cider producer 2017

When discussing how human labor and interactions with the natural world shaped the taste of cider in their region, cider producers noted the effect of variables such as farmers' irrigation practices or facilitating wild fermentation from naturally occurring yeasts that were typical of the region. Some cidermakers emphasized the artfulness involved in blending cider from locally available varieties of apples to achieve a desired taste, which was partly informed by their perceptions of local consumers' palates.

When it came to hired labor, however, orchard workers were not represented as part of the conception of terroir or marketed as part of the human faces associated with specific agrarian places. In many instances, the orchards of craft cideries were too small to require hired workers. Such cideries sourced most of their apples from larger-scale local commercial farms. This created a marked gap in cidermakers' awareness of the detailed conditions of the farm workers who produced their apples. As one cidermaker from an urban cidery explained,

> We're working with orchards who we trust are making the right decisions on that file. But obviously we don't have full insight and understanding of that. I think that anybody we work with at the very least is complying with labor laws and so on. I would hate to hear – and I hope I never hear – that an orchardist that we bought apples from is doing something sketchy on that file.
>
> (Interview with BC cider producer 2017)

His comments simultaneously underscore a concern for the well-being of workers who produce raw goods and how commodity chain dynamics in cider production obscure a full picture of labor conditions in primary production.

To a large extent, apples that eventually become part of premium, place-based craft cider are produced on orchards that are entirely separate from the cidery. Even if a cidery wished to distribute premiums from GI to hired workers on other farms, the mechanisms to pursue this could be inordinately complex. A cidery purchasing off-site apples might apply an additional layer of oversight such as third-party fair-trade certification or attempt to procure from unionized farms, which are rare or nonexistent in many apple-producing regions. A cidery producing all of its own apples could conceivably distribute GI premiums to on-site workers. However, this scenario would be highly unlikely given factors such as market competition between cider owners and rules governing wages for migrant farm workers.

Beyond the logistical challenges, there are intractable cultural barriers and a lack of political will to redistribute the benefits of a place-based North American cider marketing arrangement to hired apple workers. During my fieldwork, I frequently observed the contrast between rustic, charming cider-tasting rooms that displayed a romanticized version of agrarianism and the interior of orchard workers' employer-provided homes. It was not unusual for employers to vigorously screen and control the movements of visitors like me during workers' free time. While some homes were spacious and well maintained, others were tiny, dilapidated and heavily infested with rodents. I was struck by how contexts of relative poverty and precariousness could coexist alongside sites of elite cultural consumption premised on the idea of transparency through agritourism. Addressing long-standing, systemic and politically rooted inequities for farm workers does not square easily with the consumer-oriented market for an experience of agrarian nostalgia. The barriers to overcoming this contradiction through GI as a tool for emancipation and democracy could be considerable.

Discussion

The development of a market for premium, place-based craft cider appears to have enabled socioeconomic benefits for new and young farmers who, among those I interviewed, were disproportionately white. By circumventing the broader pattern in which apple orchards and processing sites have been consolidated into fewer and fewer hands, this value-added product appears to have allowed more producers to stay on the land. Given the risks posed by large cider-producing companies that are appropriating and redefining the boundaries of craft cider, formalized certifications for place-based products such as GI could act as a tool to signify distinctiveness and protect the niche market for small- and medium-scale orchard-based cidermakers, at least in the short term. For hired workers, GI does not appear to promise the same benefits. A pro-labor GI code of practice could not address systemic issues such as the militarization of the US-Mexico border and workers' deportability, legal jurisdictions in which workers cannot access workplace democracy through unionization or the violence and globalized inequity that inform people's decision to leave their countries of origin to support themselves and their families through wage labor.

In the agrarian imaginary analyzed by Besky (2014) as part of Darjeeling tea plantations, workers' situatedness in the agricultural landscape and the fetishization of their labor are crucial components of conceptualizing a place-based artisanal GI product. Despite this symbolic recognition, Besky argues that the material premiums from fair trade and GI certification of Darjeeling tea have generally not flowed into tea pickers' pockets. In the case of North American craft cider, I have highlighted barriers to distributing premiums from a hypothetical GI system to workers. Conceptions of terroir and the pursuit of GI designation are at such a budding, limited and uncertain stage that if pro-labor considerations were seen as sufficiently important, conceptions of terroir in North American craft cider could be reconfigured in a way that rewards both farmers and farm workers by inscribing labor laws into GI regulations. Such a policy innovation has not permeated migrant worker–dependent GI production in Europe. Focusing on workers in the Global South, Coombe and Malik (2018) advocate for alternatives to European models of GI that instead foreground rights-based development, self-determination and inclusivity in decision-making and the distribution of value.

Based on my fieldwork, a picture emerges of an agrarian imaginary for craft cider in which low-wage, racialized and precarious orchard workers become a problem to be normalized, as with the vignette at the beginning of this chapter, or as a factor or production that is distant and obfuscated from the final product. Thus, the attempt to make transparent certain aspects of food production through place-based marketing can simultaneously re-fetishize certain aspects of labor behind a commodity. Preliminary conceptions of craft cider's terroir in North America encompass everything from macro-level factors such as climate and geology, meso-level human farming practices and minutiae such as plant genetics and wild yeasts. Hired workers constitute the foundation of craft cider production. Yet the labor of hired workers in this case is not positioned as relevant to the nature-human relationships that define terroir. On the one hand, craft cider forms part of the renaissance of manual livelihoods and new cultural signifiers of occupational status (Ocejo 2017). On the other hand, it reinscribes old boundaries of race and class. Farm laborers' markers of social difference, particularly with regard to class and racialization, stand in marked contrast to rural imaginaries of whiteness and relative affluence within the realm of farm ownership and leisure-based agritourism (Tomic and Trumper 2016). At present, it is unclear how formalizing a place-based designation of craft cider as a terroir product would underpin socioeconomic or democratic benefits for hired farm workers. More fundamental structural barriers to emancipation and democracy for workers would require significant policy changes (e.g., Weiler 2018).

Conclusion

In this paper, I have described the emerging North American craft cider industry, in which there is heterogeneous interest among cidermakers in pursuing GI designation as a tool for industry protection and economic development. Place-based marketing already figures heavily in craft cider. Many cider producers are still in the early stages of business development, and they tend to see the development of stable enterprises and collective definitions of cider's "taste of place" in a given region as necessary precursors to the pursuit of a formal GI designation. Given this context, I have analyzed how the renaissance of place-based artisanal craft cider as a manual occupation rewards those engaged in producing raw ingredients.

Echoing Sekine's (this volume) conclusion, public policy and other institutions would be necessary for GI to meaningfully strengthen democracy and social justice for members of equity-seeking food producers. For the socioeconomic and democratic promise of place-based designations such as GI in the North American craft cider industry to flow to hired farm workers and their families, significant progressive changes would be necessary for immigration policy and labor laws, which would need to address the root causes of migration and displacement. An equitable system of GI would also ensure that workers had meaningful governance power over GI premiums, their own movement across borders and their working conditions. Finally, the North American craft cider industry would need to cultivate a more expansive vision of who belongs in specific agrarian landscapes and whose labor counts as "craft."

References

Aaron, Paul, and David Musto. 1981. "Temperance and Prohibition in America: A Historical Overview." In *Alcohol and Public Policy Beyond the Shadow of Prohibition*, edited by M. H. Moore and D. R. Gerstein. Washington, DC: National Academy Press. www.ncbi.nlm.nih.gov/books/NBK216414/#ddd00067.

Arcury, Thomas A., Joseph G. Grzywacz, Jennifer W. Talton, and Haiying Chen. 2010. "Repeated Pesticide Exposure Among North Carolina Migrant and Seasonal Farmworkers." *American Journal of Industrial Medicine* 53 (8): 802–13.

BCLDB. 2018. *Sales Agreements*. British Columbia Liquor Distribution Branch. Accessed February 16, 2019. www.bcldb.com/doing-business-ldb/sales-agreements.

Ben Hassen, Tarek, and Diane-Gabrielle Tremblay. 2016. "Labels of Origin and Terroir: The Case of the Certified Wine Label 'Québec Certified Wines' and the Protected Geographical Indication 'Ice Wine Québec'." *Journal of Wine Research* 27 (1): 41–64.

Besky, Sarah. 2014. *The Darjeeling Distinction: Labor and Justice on Fair-Trade Tea Plantations in India*. Berkeley and Los Angeles: University of California Press.

Binford, Leigh. 2013. *Tomorrow We're All Going to the Harvest: Temporary Foreign Worker Programs and Neoliberal Political Economy*. Austin, TX: University of Texas Press.

Bonanno, Alessandro, and Josefa S. B. Cavalcanti. 2014. "Introduction." In *Labor Relations in Globalized Food*, edited by A. Bonanno and J. S. B. Cavalcanti, xiii–xlix. Bingley, UK: Emerald Group Publishing Limited.

Bowen, Sarah. 2015. *Divided Spirits: Tequila, Mezcal, and the Politics of Production*. Oakland, CA: University of California Press.

Cairns, Kate, and Josée Johnston. 2015. *Food and Femininity*. London and New York: Bloomsbury Academic.

Cappeliez, Sarah. 2017. "How Well Does Terroir Travel? Illuminating Cultural Translation Using a Comparative Wine Case Study." *Poetics* 65: 24–36.

CARTV. 2014. "PGI – Québec Ice Cider: Designation Officially Reserved December 30, 2014." *cartv.gouv.qc.ca*. Accessed April 27, 2018. www.cartv.gouv.qc.ca/en/reserved-designa tion-pgi-quebec-ice-cider.

Coombe, Rosemary J., and S. A. Malik. 2018. "Transforming the Work of Geographical Indications to Decolonize Racialized Labor and Support Agroecology." *UC Irvine Law Review* 8: 363–412.

Coplen, Amy K. 2018. "The Labor Between Farm and Table: Cultivating an Urban Political Ecology of Agrifood for the 21st Century." *Geography Compass* 8 (3): e12370–12.

Eakin, Hallie, Stuart Sweeney, Amy M. Lerner, Kirsten Appendini, Hugo Perales, Douglas G. Steigerwald, Candida F. Dewes, Frank Davenport, and Julia C. Bausch. 2018. "Agricultural Change and Resilience: Agricultural Policy, Climate Trends and Market Integration in the Mexican Maize System." *Anthropocene* 23: 43–52.

Fabien-Ouellet, Nicolas, and David S. Conner. 2018. "The Identity Crisis of Hard Cider." *Journal of Food Research* 7 (2): 54–14.

Fortin, Marcel A. J. 2000. "Popular Culture and Public Drinking in Eighteenth-Century New France: Louisbourg's Taverns and Inns, 1713–1758." Master's thesis, Department of History, University of British Columbia. https://open.library.ubc.ca/cIRcle/collections/ubctheses/831/items/1.0099553.

Futrell, Susan. 2017. *Good Apples: Behind Every Bite*. Iowa City: University of Iowa Press.

Gray, Margaret. 2014. *Labor and the Locavore: The Making of a Comprehensive Food Ethic*. Berkeley and Los Angeles: University of California Press.

Guthman, Julie. 2014. *Agrarian Dreams: The Paradox of Organic Farming in California*. 2nd ed. Berkeley: University of California Press.

Hennebry, Jenna, Kerry Preibisch, and Janet McLaughlin. 2010. *Health Across Borders: Health Status, Risks and Care Among Transnational Migrant Farm Workers in Ontario*. Toronto, ON: CERIS Ontario Metropolis Centre.

Jones, Brad. 2016. "The Butcher, Vintner, and Cidermaker: Crafting Good Food in Contemporary America." *Cuizine: The Journal of Canadian Food Cultures* 6 (1): 1–13.

Jordan, Jennifer. 2016. *Drinking Revolution, Drinking in Place: Craft Beer, Hard Cider and the Making of North American Landscapes*, 1–9. Dublin: Dublin Institute of Technology. https://arrow.dit.ie/dgs/2016/June1/12/.

Kerrigan, William. 2012. *Johnny Appleseed and the American Orchard: A Cultural History*. Baltimore, MD: Johns Hopkins University Press.

Le Goffic, Caroline, and Andrea Zappalaglio. 2017. "The Role Played by the US Government in Protecting Geographical Indications." *World Development* 98: 35–44.

McLaughlin, Janet, Jenna Hennebry, and Ted Haines. 2014. "Paper Versus Practice: Occupational Health and Safety Protections and Realities for Temporary Foreign Agricultural Workers in Ontario." *Perspectives interdisciplinaires sur le travail et la santé* 16 (2): 1–20.

Minkoff-Zern, Laura-Anne. 2014. "Hunger Amidst Plenty: Farmworker Food Insecurity and Coping Strategies in California." *Local Environment* 19 (2): 204–19.

Ocejo, Richard E. 2017. *Masters of Craft: Old Jobs in the New Urban Economy*. Princeton, NJ and Oxford, UK: Princeton University Press.

Otero, Gerardo. 2011. "Neoliberal Globalization, NAFTA, and Migration: Mexico's Loss of Food and Labor Sovereignty." *Journal of Poverty* 15 (4): 384–402.

Paz Ramirez, Adriana, and Jennifer J. Chun. 2016. "Struggling Against History: Migrant Farmworker Organizing in British Columbia." In *Unfree Labour Struggles of Migrant and Immigrant Workers in Canada*, edited by A. Choudry and A. Smith. Oakland, CA: PM Press.

StatCan. 2017a. "Data Tables: 2016 Census: Employment Income Statistics." *Statistics Canada*. Accessed February 16, 2019. https://www12.statcan.gc.ca/census-recensement/2016/dp-pd/dt-td/Index-eng.cfm.

StatCan. 2017b. "A Portrait of a 21st Century Agricultural Operation." *Statistics Canada*. Accessed February 15, 2019. www.statcan.gc.ca/pub/95-640-x/2016001/article/14811-eng.htm.

Tomic, Patricia, and Ricardo Trumper. 2016. "Laboring in the 'Four-Season Paradise': Workers and Agriculture in the Okanagan Valley." In *Farm Workers in Western Canada: Injustices and Activism*, edited by S. A. McDonald and B. Barnetson, 187–205. Edmonton, AB: University of Alberta Press.

USDA. 2014. "Farm Demographics: U.S. Farmers by Gender, Age, Race, Ethnicity, and More." *USDA Census of Agriculture* 1–4. Accessed April 18, 2018. www.agcensus.usda.gov/Publications/2012/Online_Resources/Highlights/Farm_Demographics/#average_age.

———. 1860. "1860 Census Publications: Agriculture of the United States in 1860." *USDA Census of Agriculture Historical Archive*. Accessed April 3, 2018. http://agcensus.mannlib.cornell.edu/AgCensus/censusParts.do?year=1860.

USDOL. 2016. "Findings from the National Agricultural Workers Survey (NAWS) 2013–2014: A Demographic and Employment Profile of United States Farmworkers." *US Department of Labor* 1–89. Accessed September 24, 2018. www.doleta.gov/agworker/pdf/NAWS_Research_Report_12_Final_508_Compliant.pdf.

Watkin, Renata. 2018. "Placing Canadian Geographical Indications on the Map." *Intellectual Property Journal* 30 (2): 271–308.

Weiler, Anelyse M. 2018. "A Food Policy for Canada, But Not Just for Canadians: Reaping Justice for Migrant Farm Workers." *Canadian Food Studies* 5 (3): 279–84.

Weiler, Anelyse M., Janet McLaughlin, and Donald C. Cole. 2017. "Food Security at Whose Expense? A Critique of the Canadian Temporary Farm Labour Migration Regime and Proposals for Change." *International Migration* 55 (4): 48–63.

Weiler, Anelyse M., Gerardo Otero, and Hannah Wittman. 2016. "Rock Stars and Bad Apples: Moral Economies of Alternative Food Networks and Precarious Farm Work Regimes." *Antipode* 48 (4): 1–23.

12 The potential role of geographical indication in supporting Indigenous communities in Canada

Donna Appavoo and Monika Korzun

Introduction

In a recent report by the UN Special Rapporteur on the Right to Food, the current level of food insecurity among Indigenous communities in Canada is alarming (De Schutter 2012). The prevalence of food insecurity among some Indigenous communities in Canada is more than four times the rate in the rest of the country (Council of Canadian Academies 2014). A high level of food insecurity among Indigenous communities in Canada not only impacts their physical, mental and spiritual health but also puts at risk their ability to continue living on their traditional lands, preserve their ways of life and pass on their history, heritage, culture and knowledge to future generations. Programs attempting to address food-related issues among Indigenous populations to date have not taken into account the unique issues these communities face (Fieldhouse and Thompson 2012. Issues arising from ongoing colonialism, like land access, environmental dispossession and erosion of identity, culture and knowledge, are largely ignored in program development. As the awareness of Indigenous peoples' history and experiences continue to rise, the need to protect knowledge and ways of life, including food systems, is increasingly recognized as a vital component of improving the complex circumstances of Indigenous populations in Canada.

This chapter aims to explore the opportunities and challenges of implementing protected market access for Indigenous foods using geographical indication (GI) schemes. Federal and provincial support for commercially promoting specific products produced uniquely within Indigenous communities in Canada is nonexistent. Little research has been dedicated to exploring the potential of commercial practices as a means of addressing issues among Indigenous population in Canada (see Islam and Berkes 2016; Lowitt et al. 2018). GIs provide intellectual property protection and give permission to producers to use a registered name that is linked to a specified area. Therefore, only specific producers, usually within a narrowly defined geographical area and following specific processes, recipes and rules are given permission to use a designated name for their products. The registered names cannot then be used in marketing comparable products from other regions.

Although GI is an instrument to promote products commercially, the process can yield intangible benefits such as maintaining history, traditions and local knowledge and social benefits such as increased rural employment, agritourism and rural alliances. There are numerous parallels between the various issues that Indigenous communities are facing and the potential benefits of GIs. As mentioned previously, Indigenous communities face significant barriers to preserving their history, heritage and local knowledge. In addition to helping overcome these barriers, it is worth asking whether there is potential for GIs to reduce the rate of food insecurity, uplift Indigenous foods and culture, empower communities and educate the public about Indigenous history. Caution ought to be exercised when exploring these questions as GIs are embedded in a capitalist framework that when deployed in certain ways, also has the potential to marginalize or even disadvantage vulnerable actors (Weiler, this volume) while promoting uneven modernization (Niederle Wilkinson and Mascarenhas; Renard and Arista, this volume) and/or producing and reproducing economic power imbalances (Sekine, this volume). Further worrying questions arise when thinking about aligning GI with the (contemporary) geography and movement of Indigenous peoples, the existing governance arrangements, the establishment of ownership claims to agricultural products and foodstuffs and the potential for misappropriating Indigenous knowledge for profit.

The chapter will first outline a brief history of Indigenous populations and government relations in Canada and provide a brief outline of GIs. The ensuing discussion section of the chapter will explore the potential of GIs in Indigenous communities in Canada, contouring the various benefits of GIs but also exploring some assumptions and criticisms of developing a GI scheme in such a unique context. This analysis is explorative, and as such, it aims to illuminate the opportunities and challenges of potentially implementing GIs for the benefit of Indigenous communities in Canada.

Literature review

Indigenous communities in Canada

This section provides an overview of the policies and processes that impact the food systems of Indigenous communities in Canada today. While these processes are linked to a colonial history, the effects continue to be felt today. The impacts also vary by community, as there is considerable diversity among Indigenous communities in Canada, including more than six hundred Indigenous communities that speak over seventy different languages with specific historical cultural affiliations and traditions (AFN 2016). As a nation-state, the country of Canada is home to more than 1.4 million Indigenous peoples (Statistics Canada 2017). Indigenous populations in Canada comprise three main groups: First Nations, Inuit and Métis, each with its own political structures. Since the confederation of Canada as a nation-state in 1867, Indigenous peoples as a whole have been subject to various programs and policies that aimed to assimilate them to the culture and belief systems of the white settlers (Truth and Reconciliation Commission of Canada 2015).

Part of the mandate of assimilating Indigenous populations was to alter how food was accessed, prepared and consumed (Truth and Reconciliation Commission of Canada 2015; Mosby and Galloway 2017). One recently documented mechanism of cultural genocide was the Indian residential school system in place between 1867 and 1996. This system has received widespread public attention since the publication of the Truth and Reconciliation Commission report in 2015 and the initiation of government processes aimed at redressing past injustices (Indigenous and Northern Affairs Canada 2018). In the residential school system, children as young as four years old were taken from their families for the school year, often during key food harvest seasons (Truth and Reconciliation Commission of Canada 2015). The children were relocated to residential schools. At the schools, Indigenous children were penalized for speaking their native language and were forced to change the clothes that they wore, with important symbolic characteristics such as long hair among boys being removed. Foods served at the schools were often nutritionally inadequate and did not resemble those from "home" in how they were prepared or in their ingredients. Access to meats such as seal and moose, staples of some Indigenous diets at the time, were not provided. Not only were there negative health and social impacts, but also culturally embedded knowledge and related food skills for harvesting, gathering and hunting practices were largely lost. These and other policies have had, and continue to have, enormous negative impacts on Indigenous populations in several ways, particularly access to land and resources that support their systems of food acquisition and sharing.

Contemporary relations between the Canadian government and Indigenous communities continue to be shaped by the Indian Act, the primary document that outlines the treaties and the relationship between Indigenous peoples and Canada. First established at the time of confederation in 1867 to define the rights and responsibilities of Canada as a nation with regard to Indigenous First Nations, this collection of treaties has continued to define relationships to the present day. Principally, treaties were Indigenous land surrenders on a huge scale with small "reserve" tracts of land set aside for the exclusive use of the Indian band. Band members possess the right to live on reserve lands, and band administrative and political structures are frequently located there. This land area now makes up approximately 0.2 percent of the Canadian land mass but is home to about 50 percent of the people who identify as Indigenous in Canada (Indigenous and Northern Affairs Canada 2015). Treaties in place under the Indian Act place direct restriction on food harvesting on reserves. For example, food that is obtained on reservation land, either through growing, harvesting, gathering or hunting cannot be sold in any commercial market. Not only are foods harvested on reservations prohibited from being sold, but they are also subject to food regulations that prohibit them from being distributed, even at no cost, at public institutions such as hospitals or schools that are also on the reserve. Although there is progress toward improved land access and governance for and by Indigenous people, such as the Nunavut Land Claims Agreement Act in 1999, the process requires significant time and resources (Indigenous and Northern Affairs Canada 2015).

Modern conservation efforts can also impact Indigenous practices and ways of knowing (Pictou 2018). For instance, government policies may forbid Indigenous populations from entering traditional lands and waters not specified in treaties when these lands and waters are designated for specific conservation purposes. Pictou (2018), gives the example of ocean Marine Protected Areas (MPAs) in ocean waters off of the east coast of Canada where fish harvest is prohibited. These MPAs are often established on the best fishing grounds with little or no input from Indigenous communities, yet effectively displace Indigenous people from ancestral fishing grounds. Understandably, confidence in the capacity of the nation-state to support the reclamation and celebration of Indigenous food system should therefore be met with skepticism. Rebuilding this trust will require specific action with meaningful participation from the Indigenous community.

Geographical indications

Although GIs started in Europe as a scheme to help valorize, protect and market specific products, the concept has evolved through its extension into new regions, and the diversity of its implementation is now largely defined by new third-country policy innovations (Feuer, this volume). The European Union defines GIs as "a distinctive sign used to identify a product as originating in the territory of a particular country, region or locality where its quality, reputation or other characteristics is linked to its geographical origin" (European Commission 2013, 1). GIs contain names of agricultural products and foodstuffs, including dairy products, meats, fish, fruits and vegetables, baked goods, coffee and condiments. The names and logos of registered products can be used only in the way that have been defined by the producers association for each GI.

Canada has historically viewed GIs as a form of protectionism that interferes with the global commodity market. With the advent of the ratification of the Canada-EU Comprehensive Economic and Trade Agreement (CETA), Canada's perspective on GIs may change. Currently, Canada does not have any federal GI schemes protecting products grown or produced in Canada. In 2006, the province of Quebec implemented legislation for agri-food GIs. Although this is the most developed GI scheme in Canada, it is largely in its infancy (Cadogan 2018; Watkin 2018). Several products have been registered to date, including sweet corn, cheese, lamb, ice cider and ice wine (Cadogan 2018). The Quebec-based scheme is recognized only within the province of Quebec and does not receive any federal support. Further, upon the signing of CETA, the Canadian government did not recommend including any of the Quebec-based GIs in the EU system. Lack of interest in GIs among producers was cited as the main reason. However, there is no indication that any consultations on this topic were done with any producers, suggesting other motivations for this decision. After reviewing several transcripts from consultations, meetings and discussions about CETA with various actors, the topic of GIs is discussed only in the capacity of accepting EU GIs into the Canadian intellectual property system and how EU GIs will impact the Canadian dairy industry. There is no suggestion of including Quebec GIs in the EU system or adopting GIs as a potential scheme for Canadian producers.

In a recent assessment of the current policy landscape, Cadogan (2018) claims that while CETA has opened new opportunities in Canada for GI schemes, there is a lack of understanding and infrastructure about potentially implementing GIs (Cadogan 2018). Watkin (2018) suggests the complex constitutional framework, the lack of a uniform and comprehensive legal framework and the lack of cooperation between provincial and federal governments make it difficult for a GI program to be implemented by any jurisdiction in Canada. Finally, research about consumer awareness of GIs in Canada is low. The capacity for GI schemes to yield benefits to producers therefore hinges on broad education among decision makers, as well as the private sector and the public. Cadogan (2018) suggests that increasing consumer awareness will not only foster the growth of GI businesses but may also decrease fraudulent representation of products in domestic markets.

Although GIs have an essentially commercial orientation, it is generally recognized that GIs can aid in addressing various other issues faced by vulnerable populations, including preserving not only the specific nutrient and sensory properties of a foodstuff but also the traditions, knowledge and environments that are unique to a geographical production area (Bramley, Bienabe and Kirsten 2009). For some developing countries, the adoption of GI is considered an efficient way of using the existing intellectual property framework to promote agricultural products within a globalized world and to ensure that the names of products will not be misappropriated (Durand and Fournier 2017). While the preservation of the unique characteristics of foodstuffs and traditions is generally assessed as yielding benefits to producers and consumers alike, there are also analyses that draw attention to cases in which GIs have served to re-entrench or even deepen uneven power dynamics in a food value chain (see Sekine, this volume). And yet GIs remain a hopeful solution in many contexts for marginalized rural communities that cannot compete on a large agricultural level while helping to diversify agricultural areas that are already partaking in industrial production (Grote 2009; Dagne 2010; Durand and Fournier 2017).

The main mechanism of GIs is to provide legal protection to producers. For producers, GIs restrict the use of the geographical names and help protect the reputation and the knowledge that complements production in order to create transparency while protecting products from fraud and forgery. For consumers, GIs provide traceability and ensure compliance with the determined code of practice (CoP) (Medeiros, Passador and Passador 2016). GIs provide more information about products and the process of production than non-registered products do (Espejel, Fandos and Flavian 2007; Medeiros, Passador and Passador 2016). Registering the product and using the logos allows producers to send a message to customers about the specific characteristics and origins of the products. This transparency and educational mechanism embedded in the GI system and logo presents an opportunity for decreasing the relational distance between producer and consumer groups across cultural and geographical distances.

Several authors have also identified GIs as a spark for collective action. Durand and Fournier (2017) claim that the product itself and the protection of the product from misappropriation and misrepresentation demand collective action from the stakeholders. Collective action is most feasible when the product is more specified, because the producers understand the uniqueness of the product. Collective action can also be strengthened when the intention of the GI product is to protect a specific community and when the production process depends on communal goals (Allaire and Sylvander 1997). Collective action and partnerships are also developed to provide services such as agritourism, catering services or crafts. GIs have the potential to unite local actors, empower local organizations and ultimately strengthen rural communities (Allaire and Sylvander 1997; Durand and Fournier 2017; Cadogan 2018). GIs can also contribute to the appreciation of the product among the producers themselves, encouraging them to foster pride in their spaces and ways of life.

Discussion

Based on the literature review above, this section will examine the unique opportunities and challenges that may arise with the potential development of a GI scheme as a support tool for Indigenous communities in Canada. The discussion is divided into four sections. These sections outline key issues that may be unique to potential GI schemes in Indigenous communities. The sections are as follows: local knowledge, land access, market access and legal protections.

Local knowledge

GI schemes are often promoted as a process of protecting, validating and even valorizing local knowledge. Importantly, tradition or local knowledge does not exclude innovation (Bertoni, Calamari, and Maianti 2001). There are many examples of when techniques involved in traditional technology were at one point considered modern. Many technological developments from the end of the 18th century, such as centrifuge, pasteurization and lactobacillus cultures were defended despite their relative modernity and are now considered part of traditional cheesemaking. Also, despite technological developments and the enlargement of dairies, many components of cheesemaking, such as using raw milk, natural culture and slow salting and ripening have remained. Bertoni, Calamari and Maianti (2001) suggest that the success of cheeses under the EU GI scheme are in part a result of the combination of modern technology and tradition, with the goal of preparing the product for market demand while minimizing compromises to the authenticity and originality of the product.

Despite the technological advancements, it is this context of specific knowledge that guides the development of GIs, allows for differentiation and distinguishes cheeses from one another (Bertoni, Calamari and Maianti 2001).

Dagne (2010) claims that GIs have the potential to protect local knowledge and ways of life while empowering the producers. Hayes, Lence and Stoppa (2004) also found that local producers' pride in their knowledge of particular products increases when their products are registered and promoted via the GI schemes. The pride in their knowledge and products translates to the producers paying closer attention to the integrity of the products and discouraging compromises in traditional production practices. GI schemes are therefore valuable not only because of their protection of traditional production practices but also because they can contribute to continued economic viability, raising consumer awareness and the sustainability of rural communities. As discussed in this book's introductory chapter (see Bonanno, Sekine and Feuer, this volume), numerous studies have shown that the programs are significant components in regenerating rural communities, because the traditional and local foodstuffs are produced by using local plant varieties, support rural diversity and social cohesion and provide an income opportunity for local producers, processors and other services. This encourages established and young community members to stay in rural areas, functionally protecting the transmission of knowledge across generations (Van der Ploeg 2002; Folkeson 2005).

Although traditions and local knowledge are not a prerequisite for all GIs, many products do incorporate some aspect of this with the goal of maintaining history, heritage or local knowledge or a specific region. While every region has a unique history, heritage and local knowledge, Indigenous populations are inherently distinct by virtue of their being openly and aggressively cast as retrograde. The local knowledge and ways of life of Indigenous peoples have been used to justify horrific cultural and ethnic genocide, including family invasions such as residential schools, community relocations to unfamiliar and inhospitable conditions and the dispossession of productive food resources (Mosby and Galloway 2017). Terms which are being valorized now, such as traditional and local, were used to label the dietary practices of Indigenous populations as backward. For generations, local knowledge and traditional ways of sustenance of Indigenous populations in Canada have carried a negative connotation, which has even to some extent been perversely internalized by Indigenous populations and the Canadian public. Therefore, expending extra effort to recognize this unique experience and understanding when discussing how to maintain or revitalize heritage or traditions is important. Here, GIs provide a promising combination of measures to manage the modern realities of such heritage preservation. This is because, on the one hand, Indigenous populations want to revive their knowledge and way of life while, on the other hand, demonstrating that their communities can evolve and be relevant to modern ways of life. Whereas valorizing local knowledge and traditional ways of life can be associated with a great deal of pride for many producers in Europe (Hayes, Lence and Stoppa 2004), the use of certain ingredients or the participation in various acts, such as hunting moose, can bring forward a complex variety of emotions, memories and associations among Indigenous people in Canada.

Indigenous knowledge keepers also believe that traditions and local knowledge can evolve. As King (2018) explains, vernacular knowledge arose based on a preexisting suitability that demanded flexibility, mobility and broad-scale evolution over time. This stands in direct counterpoint to colonial attitudes dismissive of knowledges held in Indigenous communities (Matties 2016). Indigenous knowledge holders caution that the word *traditional* may too easily be perceived as fixed in a specific time and place and be difficult to make relevant to the current state of Indigenous communities. Further, using the word *traditional* to describe Indigenous populations may also imply that Indigenous communities cannot evolve or adapt to changing socioecological circumstances. For example, despite many stereotypes in Canada, Indigenous peoples have been using techniques that would have been unavailable to previous generations, such as motorized vehicles to gather or hunt for decades. In other words, traditional practices can evolve to incorporate modern technology and resources.

Indigenous knowledge is shaped by an explicit responsibility to the needs of present and future generations. For example, the Seventh Generation Principle, which has been incorporated into various sustainability models, is based on Iroquois philosophy: a decision ought to take into consideration the impact that that decision will have on seven generations into the future. Further, in recent governance arrangements yielding greater Inuit autonomy in the Nunavut territories of Canada, the wildlife act informed by Inuit principles of stewardship and respect for wildlife and the environment include explicit reference to *Avatimik Kamattiarniq*, meaning that people are stewards of the environment and must treat all of nature with respect, because humans, wildlife and their shared environment are interconnected and each person's actions and intentions toward every environmental entity have consequences for good or ill for the present and the future (Government of Nunavut 2003). Traditional and local knowledge in Indigenous communities is considered flexible because traditions and ways of knowing largely depend on the relationship between people and land, which evolve as people and environments change.

Land access

As in many cultures, food in Indigenous communities plays a key role in maintaining and creating identity, upholding traditions and celebrating various stages of individual and community life (Leblanc and Burnett 2017). Among Inuit, for example, the sacredness of food underpins the belief that the foods one eats become a part of oneself, thus representing the sacrifice that an animal makes in this intimate connection (Arnaquq-Baril 2016; Martin and Amos 2017). The special relationships with food stem largely from the close relationship that Indigenous communities have with land and water. Indigenous knowledge keepers draw attention to the importance of the land in Indigenous cultures. The land-based practices associated with production, harvesting and consumption are embedded in the Indigenous cultural understanding of food as sacred and integral to spiritual and physical well-being (Martin and Amos 2017; Morrison 2017; Pictou 2018). As indicated in the overview of government policies under the Indian Act in Canada that relate to Indigenous access to land and resources, legal access to ancestral territories, particularly hunting and fishing grounds, remains unresolved.

Perpetuation of the discourse of conservation through a lens of non-Indigenous cultural norms and values also impacts Indigenous food systems (Arnaquq-Baril 2016). This is a space where colonial and Eurocentric perceptions run counter to the Indigenous worldview about achieving equilibrium or sustainability of nature and land. Eurocentric ideologies negatively impact public opinion on traditional hunting practices such as the seal hunt. Environmental dispossession, the impact of climate change and chemical contamination, not to mention the commercial fishing industry, are the result of actions not under the control of Indigenous people, but which nonetheless reduce the availability of Indigenous foods (Doolan, Appavoo and Kuhnlein 1991; Tobias and Richmond 2014). The potential development of GI schemes in Indigenous communities will rely largely on specific land use, water use and climate conditions. The very existence of GIs may be threatened if the environment is negatively impacted, put at risk or prohibited from use. GIs have the ability to determine the scale of production systems and in many cases can prevent the use of tools and technology prominent in the industrial food system that unsustainability deplete resources. This may discourage industrialization of agricultural production and processing, ultimately putting less pressure on the ecosystem aligning with the connection to and respect for the land and to the needs of future generations that is inherent in Indigenous food systems. The centrality of responsibility to the land in Indigenous approaches can in turn be a valuable lens to identify features that could strengthen the GI process, specifically prioritizing potential benefits to ecosystems and land integrity in GI scheme development.

Market access

Due to the negative stereotypes of Indigenous people in Canada, a false perception that Indigenous communities do not partake or do not wish to partake in the market economy has persisted. Indigenous communities' involvement in trade and market spaces have, more accurately, been erased through colonial processes. Berkes (2018) refers to the concept of the "mythology of Indigeneity." This mythology, along with labeling Indigenous communities as primitive and backward, has largely restricted their participation in market spaces in Canada. This is an immensely problematic and consequential issue for maintaining the integrity of food systems over generations (Berkes 2018; Palmateer 2018). This myth can act as a barrier to adapting traditional systems to contemporary contexts, thereby limiting the development of sustainable local economies in which Indigenous people can make a living in the globalized world (Berkes 2018). Here is where mechanisms such as GI appear promising.

The "mythology of Indigeneity" is not only common among the general Canadian population, but is physically inscribed in numerous trading agreements of which the government of Canada is part. Section 35 of Canada's Constitution Act of 1982 requires the government of Canada to consult with Indigenous populations when developing international trade agreements such as the North American Free Trade Agreement (NAFTA) or CETA. However, none of the trading agreements in which the government of Canada is a signatory have been developed with meaningful consultation of Indigenous peoples (Schwartz 2017). Although the government of Canada does not abide by Section 35 in practice, during trading agreement negotiations, Section 35 has been viewed suspiciously as a potential obstacle or a trade barrier to both the government of Canada and potential trading partners (Schwartz 2017). This lack of inclusion for Indigenous populations in decision-making and the lack of recognition of Section 35 stems from the "mythology of Indigeneity," that Indigenous peoples anyway do not engage in trade or the market economy. As Turner and Loewen (1998, 49) claim, "Archaeological and historical records show that they developed extensive and sophisticated trading networks and institutions for thousands of years. However, the pervasive nature of Indigenous trade, and its many implications both before and after contact, have rarely been explicitly considered." Even though this idea has been explored for some time and documentation of the "mythology of Indigeneity" is more recently discredited in a range of literatures, this myth continues to persist in spaces that can impact Indigenous access to market.

To improve conditions related to food security, communities require not just physical access to foods that supports health but also capacity to be able to support and maintain access to market (Kuhnlein, Erasmus and Spigelski 2009; Kuhnlein et al. 2013). To achieve this, the "mythology of Indigeneity" ought to be broken down. Further, special conditions, programs or funding schemes that encourage the participation of Indigenous peoples, such as GIs that encode Indigenous practices, ought to be created. Greater participation in markets for Indigenous communities would not only allow those practices to contribute to present household provisioning but also allow for the expansion of those practices. For example, based on numerous interviews with Indigenous fishers in the Great Lakes, Lowitt et al. (2018) conclude that Indigenous fishers feel marginalized in the current market economy and thereby have insufficient resources to maintain their equipment and apprentice the youth in fishing practices. Although trade and participation in markets is part of Indigenous culture and heritage, much of that history has been erased through colonial processes. Reviving the part of the culture that has been erased through colonialism, trade and market literacy among the Indigenous population, especially the youth, is vital.

Some products in the market place may yield ceremonial or spiritual value for Indigenous peoples. Along with the appropriation of lands and territories, Indigenous populations have also been subject to the appropriation of their food knowledge and traditions. For example, maple syrup is considered an iconic "Canadian" product, protected by specific provisions for producers within the contemporary system of production. However, there is no provision for protecting the specific places of production and production modalities for Anishinabek and other Indigenous nations for whom maple syrup holds specific cultural and spiritual meaning that is well documented in Indigenous storytelling (Banning 2018; Simpson 2014). In addition, like with all GI schemes, producers will have to invest financially to partake in the program. It is in these areas that the government ought to intervene. The government must not micromanage Indigenous populations and lose sight of the goal of GIs, which would be to ultimately strengthen these communities' capacity to respectfully bring recognition and protection to their food heritage (Cadogan 2018). Governments ought to develop frameworks that provide more decision-making power to Indigenous communities over their land, access to resources, conservation areas and gathering and hunting practices (Pictou, 2018). Since the federal government has yet to initiate any schemes to help protect Indigenous communities through market initiatives, this presents an opportunity for provincial and local governments to develop initiatives that bring GIs to the global stage. Regional and local governments may be better equipped with the needs and wants of Indigenous communities and how to facilitate the process in a considerate and sensitive manner. Provinces, territories and local governments can help communities with developing strategies, facilitating consultations and providing financial aid in establishing the GI schemes. The federal government can develop legal frameworks to help protect the market and limit fraud and infringement.

Legal protections

The standards and processes by which a GI product is to be produced is often determined by social, cultural and historical traditions, as well as the unique physical, chemical and flavor characteristics arising from a specific ecosystem (Medeiros, Passador and Passador 2016). Thus, local knowledge that has been passed down from generation to generation is integral to the development process of GIs. To obtain the local knowledge necessary for the development of GIs, a consultation with local producers is essential. However, consultations do not equate to inclusiveness. As in the case of all GI products, there is power involved not only in the definition of tradition and history but also in the manner of their legal inscription. In a study done by Durand and Fournier (2017) on GIs in Vietnam and Indonesia, the process of establishing and defining codes of practices (CoPs) of several GIs lacked the contribution of local knowledge. Local and traditional producers were not included in the initial discussions or in the "team of experts" that were developed to monitor GI processes and provide technical assistance. In the case of Mèo Van Mint Honey in Vietnam, the CoP was developed to intentionally exclude traditional producers, the H'Mong people, who extract honey from tree trunks and capture wild beehives to ensure their survival throughout winter, rather than using wooden hives. This case illustrates the challenges of mediating preexisting power relations within a mixed (Indigenous and non-Indigenous) producer group.

In the potential development of GIs as a tool of innovation or reaching new markets, it is important to achieve consensus and collaboration rather than to simply organize consultations (Palmateer 2018). Consultations do not guarantee representation of all stakeholders or the representation of their opinions, and experiences with Indigenous peoples in Canada are often token in nature. As already mentioned, this is apparent at the macro level as well, in which the government of Canada has ignored its constitutional requirement to partake in meaningful consultations with Indigenous communities concerning free trade agreements. Palmateer (2018) has pointed out numerous policies and programs where collaboration was promised to Indigenous peoples, but those promises were ultimately not kept. This unwillingness to engage with Indigenous peoples is unfortunate: Bicker Sillitoe and Pottier (2004) suggest that projects that harness local knowledge via a collaborative process with diverse local actors gain many advantages over projects that do not.

Durand and Fournier (2017) conclude that top-down approaches to GIs are not successful in protecting traditional knowledge and vulnerable populations. More diverse stakeholder involvement in GI development is predictive of outcomes that yield benefit across this range of stakeholders, but many examples in this volume (Feuer; Niederle, Wilkinson, and Mascarenhas; Nizam – all this volume) demonstrate that GIs are not often institutionalized in participatory ways in new countries. Although the federal government has not yet supported any GI schemes, and most provincial and territorial governments have demonstrated little interest in developing schemes, local governments may be better equipped to initiate and implement such a program. The involvement of local governments can result in greater support for local producers and producers from vulnerable populations, giving authority to local governments to facilitate GI product registration. This may improve power distribution to local authorities, who are familiar with local customs, traditions and knowledge systems. For example, in Canada, the first GI products to be registered were all in communities in the province of Quebec under specific legislation that supports an integrated development process (Watkin 2018). Local governments act as facilitators between producers and experts and can negotiate the necessary conditions for producers to consistently achieve GI stipulations. Most importantly, local governments are more likely to ensure a participatory approach by directly engaging with producers and establishing with them the parameters of GIs. Further, local governments can identify important supply chains that can aid in the production and distribution of the GI products.

Conclusion

The intention of this chapter was to explore the idea of whether GIs can potentially be used as a tool for supporting the respectful valorization and commercialization of Indigenous products in Canada – and thereby reengage Indigenous peoples in trade while protecting traditions. Indigenous peoples are facing a wide variety of issues associated with their food system. Food insecurity, health problems, loss of traditions and local knowledge, environmental dispossession and issues of land access are significant challenges to establishing a sustainable and culturally appropriate food system. The colonial legacy of weakened food systems and deficient community infrastructure has left these communities particularly vulnerable. Although many policies and programs have been implemented in Canada to support Indigenous communities, little research explores market solutions.

GI schemes have been successful in supporting community revitalization in cases from a variety of geographies and governance structures. However, as many examples in this volume show, the model has not been assessed as supportive of community-level producer groups in all contexts. Rather than a process that empowers local actors and subordinate groups, GI could be an instrument that hampers democratic forms of development. To improve the likelihood of outcomes that would support Indigenous producer groups, GI should be grounded in local knowledge systems and apply governance structures that are built to address particularities of market access and land access of socially, politically and economically vulnerable people.

References

Allaire, Gilles, and Bertil Sylvander. 1997. "Qualité specifique et systemes d'innovation territoriale." *Cahiers d'Economie et de Sociologie Rurales (CESR)* 44: 29–59.

Arnaquq-Baril, Alethea. 2016. *Angry Inuk*. National Film Board of Canada. Accessed March 12, 2017. www.nfb.ca/film/angry_inuk/.

Assembly of First Nations (AFN). 2016. *Annual Report 2015–2016*. Accessed April 27, 2017. www.afn.ca on.

Banning, Joelle. 2018. "Maple Syrup: Indigenous Governance or Business Idea?" *Anishinabke News*, March 30, 2018. http://anishinabeknews.ca/2018/03/30/maple-syrup-indigenous-governance-or-business-idea/.

Berkes, Fikret. 2018. *Sacred Ecology: Traditional Ecological Knowledge and Resource Management*. 2nd ed. Philadelphia: Taylor & Francis.

Bertoni, Guisepi, Luigi Calamari, and Maria Maianti. 2001. "Producing Specific Milks for Specialty Cheeses." *Proceedings of the Nutrition Society* 60: 231–46.

Bicker, A., P. Sillitoe, and J. Pottier. 2004. "Preface." In *Investigating Local Knowledge: New Directions, New Approaches*, edited by A. Bicker, P. Sillitoe and J. Pottier, xi–xiv. Aldershot Harts, UK: Ashgate Publishing Limited.

Bramley, Cerkia, Estelle Biénabe, and Johann Kirsten. 2009. "The Economics of Geographical Indications: Towards a Conceptual Framework for Geographical Indications in Developing Countries." In *The Economics of Intellectual Property: Suggestions for Further Research in Developing Countries and Countries with Economies in Transition*, edited by WIPO, Vol. 1: 109–141. Geneva, Switzerland: World Intellectual Property Organization (WIPO).

Cadogan, Marsha. 2018. "Making Agricultural and Food-based Geographical Indications Work in Canada." *Policy Brief* No. 125 Centre for International Governance Innovation, Waterloo.

Council of Canadian Academies. 2014. *Aboriginal Food Security in Northern Canada: An Assessment of the State of Knowledge*. Ottawa, ON: The Expert Panel on the State of Knowledge of Food Security in Northern Canada, Council of Canadian Academies.

Dagne, Teshager. 2010. "Law and Policy on Intellectual Property, Traditional Knowledge and Development: Legally Protecting Creativity and Collective Rights in Traditional Knowledge Based Agricultural Products Through Geographical Indications." *The Estey Centre Journal of International Law and Trade Policy* 11 (1): 68–117.

De Schutter, Olivier. *Special Rapporteur on the Right to Food: Visit to Canada from 6 to 16 May 2012*. Ottawa. Accessed April 27, 2017. www.srfood.org/images/stories/pdf/officialreports/201205_canadaprelim_en.pdf.

Doolan, Natalia, Donna Appavoo, and Harriet Kuhnlein. 1991. "Benefit-risk Considerations of Traditional Food Use by the (Hareskin) Dene/Metis of Fort Good Hope, N.W.T." *Arctic Medical Research*, S: 747–51.

Durand, Claire, and Stephane Fournier. 2017. "Can Geographical Indications Modernize Indonesian and Vietnamese Agriculture? Analyzing the Role of National and Local Governments and Producers' Strategies." *World Development* 98: 93–104.

Espejel, Joel, Carmina Fandos, and Carlos Flavian. 2007. "The Role of Intrinsic and Extrinsic Quality Attributes on Consumer Behaviour for Traditional Food Products." *Managing Service Quality* 17 (6): 681–701.

European Commission. 2013. "Geographical Indications." Accessed April 27, 2018. http://ec.europa.eu/trade/policy/accessing-markets/intellectual-property/geographical-indications/.

Fieldhouse, Paul, and Sarah Thompson. 2012. "Tackling Food Security Issues in Indigenous Communities in Canada: The Manitoba Experience." *Nutrition and Dietetics* 69: 217–21.

Folkeson, Carina. 2005. *Geographical Indications and Rural Development in the EU*. Lund, Sweden: University Essay, School of Economics and Management, Lund University.

Government of Nunavut. 2003. *Consolidation of Wildlife Act*. Government of Nunavut. Accessed February 8, 2019. www.canlii.org/en/nu/laws/stat/snu-2003-c-26/latest/snu-2003-c-26.html.

Grote, Ulrike. 2009. "Environmental Labelling, Protected Geographical Indications and the Interests of Developing Countries." *The Estey Centre Journal of International Law and Trade Policy* 10 (1): 94–110.

Hayes, Dermot, Sergio Lence, and Andrea Stoppa. 2004. "Farmer-owned Brands?" *Agribusiness* 20 (3): 269–85.

Indigenous and Northern Affairs Canada (INAC). 2018. *Delivering on Truth and Reconciliation Calls to Action*. Accessed August 20, 2018. www.aadnc-aandc.gc.ca/eng/1524494530110/1524494579700.

———. 2015. *Lands and Economic Development*. Accessed April 27, 2018. www.aadnc-aandc.gc.ca/eng/1100100032790/1100100032794.

Islam, Durdana, and Fikret Berkes. 2016. "Indigenous Peoples' Fisheries and Food Security: A Case from Northern Canada." *Food Security* 8 (4): 815–26.

King, Hayden. 2018. Director of the Centre for Indigenous Governance at Ryerson University Pers. Comm. March.

Kuhnlein, Harriet, Bill Erasmus, Barbara Burlingame, and Dina Spigelski. 2013. *Indigenous Peoples' Food Systems and Well-being: Interventions and Policies for Healthy Communities*. Rome: United Nations Food and Agriculture Organization.

Kuhnlein, Harriet, Bill Erasmus, and Dina Spigelski. 2009. *Indigenous Peoples' Food Systems: The Many Dimensions of Culture, Diversity and Environment for Nutrition and Health*. Rome: United Nations Food and Agriculture Organization.

Leblanc, Joseph, and Kristin Burnett. 2017. "What Happened to Indigenous Food Sovereignty in Northern Ontario? Imposed Political, Economic, Socio-ecological and Cultural Systems Changes." In *A Land Not Forgotten, Indigenous Food Security and Land-Based Practices in Northern Ontario*, edited by Courtney Mason and Michael Robideaux. Winnipeg: University of Manitoba Press.

Lowitt, Kirten, David Johnston-Weiser, Ryan Lauzon, and Gordon Hickey. 2018. "On food Security and Access to Fish in the Saugeen Ojibway Nation, Lake Huron, Canada." *Journal of Great Lakes Research* 44 (1): 174–83.

Martin, Deborah, and Margaret Amos. 2017. "What Is Good Food?" In *Critical Perspectives in Food Studies*, edited by Mustafa Koc, Jennifer Sumner and Anthony Winson. Toronto: Oxford University Press.

Matties, Zoe. 2016. "Unsettling Settler Food Movements: Food Sovereignty and Decolonization in Canada." *Cuizine* 7: 2.

Medeiros, M., C. S. Passador, and J. L. Passador. 2016. "Implications of Geographical Indications: A Comprehensive Review of Papers Listed in CAPES' Journal Database." *Innovation and Management Review* 13: 315–29.

Morrison, Dawn. 2017. "Confronting the 'Perfect Storm': How to Feed the Future." *CBC Ideas* November 24, 2017, Guelph, Canada.

Mosby, Ian, and Tracey Galloway. 2017. "'Hunger Was Never Absent'. How Residential School Diets Shaped Current Patterns of Diabetes Among Indigenous Peoples in Canada." *Canadian Medical Association Journal* 189: 630–32.

Palmateer, Pamela. 2018. "What's Up with NAFTA, Jobs and Equity?" Panel Presentation of the Jack Layton Chair, Ryerson University, Toronto, March 20, 2018.

Pictou, Sherry. 2018. "The Origins and Politics, Campaigns and Demands by the International Fisher Peoples' Movement: An Indigenous Perspective." *Third World Quarterly* 39 (7): 1411–20.

Schwartz, Risa. 2017. "Toward a Trade and Indigenous Peoples' Chapter in a Modernized NAFTA." *Policy Brief* No. 144 Centre for International Governance Innovation, Waterloo.

Simpson, Leanne. 2014. "Land as Pedagogy: Nishnaabeg Intelligence and Rebellious Transformation." *Decolonization: Indigeneity, Education and Society* 3 (3): 1–25.

Statistics Canada. 2017. *National Forest Inventory*. Standard Reports. Accessed April 27, 2018. https://cfs.nrcan.gc.ca/statsprofile/inventory/canada.

Tobias, Joshua, and Chantelle Richmond. 2014. "That Land Means Everything to us as Anishinaabe . . .: Environmental Dispossession and Resilience on the North Shore of Lake Superior." *Health and Place* 29: 26–33.

Truth and Reconciliation Commission of Canada (TRC). 2015. *Honouring the Truth, Reconciling for the Future: Summary of the Final Report of the Truth and Reconciliation Commission of Canada*. trc.ca.

Turner, Nancy, and Nancy Loewen. 1998. "The Original 'Free Trade': Exchange of Botanical Products and Associated Plant Knowledge in Northwestern North America." *Anthropologica* 40 (1): 49–70.

Van der Ploeg, J. D. 2002. "High Quality Products and Regional Specialities: A Promising Trajectory for Endogenous and Sustainable Development." Paper Presented at the in: The Future of Rural Policy, July 10–12, 2002, Siena, Italy: OECD.

Watkin, Renata. 2018. "Placing Canadian Geographical Indications on the Map." *Intellectual Property Journal* 30 (2): 271–308.

13 Conclusions

Comprehensive change and the limits and power of sectorial measures

Alessandro Bonanno, Kae Sekine and Hart N. Feuer

Introduction

Based on, and adding to, the analyses presented in this volume, this final chapter provides an overall answer to the book's research question about the emancipatory power of geographical indication (GI). It opens here with a brief summary of the overall conclusions derived from the book, which critically reflect on the potential for GI to transform and re-organize the agri-food sector in spite of broader constraints at the societal level. Partial measures, such as GI, have not demonstrated the scope of action to generate significant emancipatory changes in agri-food. This represents a structuring conclusion of this book, to which the chapters provide more detailed and policy-relevant findings about how this limited scope of GI is negotiated and discursively presented in numerous contexts. To this end, the following section stresses that while GI cannot comprehensively change the agri-food sector, it engenders a number of outcomes that are relevant for discussions and efforts about the democratization of the sector. These outcomes are synthesized into six conclusions. This section is followed by a review of the salient historical arguments on the agrarian question and the environmental sustainability question, which provide corroborating theoretical explanation for the underwhelming transformative capacity of GI in the agri-food sector. To provide a clear frame for the unfolding contents of this chapter, a brief summary of the empirical conclusions is provided here.

The first conclusion indicates that the emancipatory role of GI depends on local social relations and socioeconomic conditions. This conclusion is made explicit through the argument that the constraints represented by existing structural conditions are mitigated not by virtue of the GI policy itself but by the ways stakeholders pursue their objectives by exercising agency. This finding, which has cross-cutting relevance for the subsequent conclusions presented here, suggests that the overt focus on the GI policy should not distract from or impede parallel support for other emancipatory or democratic activities.

Turning to its internal cohesiveness, the second conclusion indicates that the success of GI depends on how GI is implemented. Specifically, the unfolding of the process of GI implementation requires forms of adaptation to existing global and local conditions. Understood as a dynamic process, it is characterized by the way numerous and complex local forces dialogue, cooperate and engage in conflict with an equally complex and numerous set of external forces. In this context, GI implementation involves a number of relevant interconnected and contradictory processes. This section of the conclusions specifically addresses the phenomena of the loss of heterogeneity, attempts to defend local identity in response to coercive standardization processes and the vulnerability of vernacular agri-food institutions.

The third conclusion addresses the ability of GI to provide protection against the unwanted consequences of the functioning of the market. The collective experience represented in this volume suggests that GI does not necessarily offer such a safeguard nor protection against the development of conditions that undervalue locally distinctive products. Indeed, the assertion that further control of markets represents a more desirable solution than business as usual is repeatedly questioned by the findings throughout this volume, demonstrating that GI generates desirable and undesirable outcomes simultaneously. Specifically, evidence that GI provides opportunities for local producers and their communities while generating positive economic results is often countered by equally convincing evidence that markets cannot be effectively controlled, that GI can be co-opted by large corporations and that its implementation can engender forms of discrimination, exclusion and neocolonialism which may undermine socioeconomic development.

This divergent outcome is comparable to the fourth conclusion, which stresses that although GI processes are a socially and politically aggregating force, they also commonly engender episodes of division and conflict. Communal action is an important tool for the empowerment of local producers and the establishment of connections between producers and supporting consumers. GI also positively affects the development of consciousness about the importance of traditional ways of producing food, the availability of quality food for consumers and cohesion of local communities and collective environmental stewardship. At the same time, however, these positive effects often create conditions that engender episodes of distrust and conflict. In this respect, GI implementation is not inherently resistant to the effects of existing power relations.

Regarding the anticipated or assumed environmental benefits, the fifth conclusion concerns the claim that GI is a tool for promoting the sustainable use of natural resources. In response to this, it is helpful to repeat the overarching conclusion in this chapter, namely that broad environmental and sustainability questions cannot be solved by sectorial measures such as GI. Furthermore, the empirical research presented in the book often rejects purported environmental claims, indicating that GI does not inherently provide impetus for the protection of the environment, biodiversity or even reduced use of agro-chemicals.

The last conclusion presented here refers to the desire of producers to remain in farming and food production despite increasing competition, a corporate dominated agri-food sector and the continuous negative effects associated with globalization and neoliberalism. Despite its limited scope of action, GI is a tool that addresses the existential question in that it allows not only producers but also consumers to follow an alternative logic that appears emancipatory and opposes the requirements of the free market. GI represents a proposal that transcends the understanding of labor activities as simply economic events and sees GI stakeholders as stewards of history, tradition and culture who valorize the human relationship to food. This aspiration, however, must coexist with processes of commodification, wherein traditions and culture are reified to implement GI and, in doing so, emerge as products to be traded with a capitalist market logic. For many GIs, success is determined by the ability to sell the largest number of products at the highest possible price. By transforming tradition and culture into products that promote sales, these processes deny the beneficial effects of social change.

These detailed conclusions about the role and impact of GI can be encapsulated in a more overarching closing statement: The emancipatory power of GI in all of its manifestations is contradictory in that it exhibits elements that promote but also hamper the democratization of agri-food and the socioeconomic conditions of stakeholders.

The sectorial relevance and contingency of geographical indication

That GI is not the panacea that can comprehensively address the problems of agri-food is recognized by the chapters of the book, covering a wide range of countries and institutional settings. Simultaneously, however, these analyses underscore that numerous important outcomes can be achieved despite GI's more limited sectorial space of action, although these may be conflicting or inadequate. These outcomes, in their emancipatory or regressive forms, are associated with the conditions that define the people and places where GI unfolds. Accordingly, *the emancipatory role of GI depends on local social relations and socioeconomic conditions*. This observation, which is readily invoked in the literature, is used both to rationalize the limitations or failings of GI and to accuse GI of being too ephemeral or inconsistent. In presenting the first conclusion of this chapter, we wish to move beyond this by framing GI as an approach to regional agri-food governance rather than pretending that it is a cohesive and tangible policy. To this end, GI operates similar to many forms of (local) governance, in which *structural conditions* (power relations; economic and class structures; political and geopolitical conditions; etc.) are met by *human agency* or the ways local actors pursue their objectives and mobilize human and social resources to achieve them. It is fitting to view it as a *contested terrain* in which contrasting forces operate to establish results. As illustrated by Nizam, the effectiveness of GI is significantly affected by the existence of empowering discursive and participatory processes (reflexive localism) or the dominance of top-down, typically authoritarian processes that limit local participation (normative localism). The impact of GI is therefore not a linear process but rather a contingent one that can be proactively or reflexively managed if planners adopt

GI as an approach rather than attempting to shoehorn advocated policy into various local contexts.

The problem of implementation

As stressed by a number of the chapters (e.g., Feuer; Piatti and Dwiartama; Hegnes and Amilien; Niederle, Wilkinson and Mascarenhas), the success of GI is *defined by how its implementation takes place* and the multifaceted dimensions that affect how the normative component (planned outcomes) translates into history (actual outcomes). This is the second conclusion that derives from the analyses presented in the volume. These multifaceted dimensions are often defined in terms of adaptation (see Hegnes and Amilien). Yet this term is not synonymous with the one-way act of conforming to an existing structure/system. Instead, it implies a dynamic and complex process in which a number of local social forces dialogue, cooperate and conflict with an equally complex collection of external forces. As this process of implementation unfolds, numerous relevant, interconnected and contradictory processes/conditions emerge, including the loss of heterogeneity, the struggle for the definition of local identity and the relative vulnerability of the relevant institutions.

As far as the loss of heterogeneity is concerned, the analyses presented in the book (e.g., Piatti and Dwiartama; Hegnes and Amilien; Niederle, Wilkinson and Mascarenhas) show that while heterogeneity is a necessary precondition for valorizing the uniqueness of candidate products, the creation of a GI demands opposing processes of conformity and standardization. This contradiction stems from prevailing requirements for characterizing GIs and rendering them transparent and amenable to international authorities and consumers. In effect, the very process of creating GIs initiates a forceful representation of the many historically based, empirical characteristics that typify each GI product. This representation materializes through single-product definitions that – as they synthetize but also accentuate and, in some cases, eliminate some of the product's characteristics – need to be compatible with preordained requirements. Accordingly, at least two fundamental and interrelated components of the GI system tend to lessen heterogeneity. First, heterogeneity is lessened by the process through which local products' tradition and cultural embeddedness are made to conform to the global (e.g., World Trade Organization) and domestic conventions (e.g., national food regulations). Second, it is lessened by the process of negotiation that emerges as necessary for the legal definition and the consequent recognition of GI status. As illustrated in the chapters, homogenization forces include not only the social process of conforming to negotiated standards but also the political logic of establishing GI policies. In the latter, GI promoters often adopt the instrumental rationality of market relations, whereby processes of homogenization are encouraged in order to eliminate differences that hinder participation in market competition. In essence, tradition and culture are created and then standardized to be almost exclusively functional to market requirements.

Part and parcel of the GI process, then, is the continuous need to respond to homogenization tendencies by mounting a defense of local identity that protects the uniqueness of GI products. These countervailing forces are inherently contradictory. As explained by Shtaltovna and Feuer, assembling a GI policy that conforms to European norms about wine and embodies local identity and tradition is a fraught task, which is only further complicated if policymakers cannot enlist all of the constituent stakeholders. Discussing the case more broadly for East Asia, Feuer stresses that the enthusiasm for GI policies rests on their promotion as an indispensable tool for touting the cultural and economic value of local agri-food. Simultaneously, however, promoters argue that the value is even further increased if directed to the global market, which often entails cross-cultural adjustment that assimilates the traits of the commodified market-based global culture and thereby lessens heterogeneity. As these chapters contend, this process problematizes not only the reevaluation of culture but also the very understanding of local identity. Moreover, and as stressed by Hegnes and Amilien and by Niederle, Wilkinson and Mascarenhas, the presentation of local identity in cross-cultural terms frequently promotes innovation in a dissonant form. Innovation for the sake of cross-cultural adjustment often alienates producers and consumers from their familiar products, contradicting the very essence of local identity. In short, the defense of local identity meant to provide a check against the inexorable homogenization tendencies of GI struggle to transcend the requirements of the market, and consequently, GI may come to repress those aspects that are not amenable to market requirements.

The actions of pertinent institutions shape GI implementation, and accordingly, the strength and orientation of these institutions brackets the defense of local identity and adaptation to global and national food systems. As demonstrated by Hegnes and Amilien in their illustration of GI in Norway, well-resourced institutions facilitate local actors' efforts to mediate between external requirements and the demands stemming from specific local conditions. The chapter by Renard and Domínguez Arista on Mexico, however, demonstrates how strong institutions can have a perverse effect if they are co-opted by corporate players. In contrast, Feuer describes how the weak GI institutions found in Cambodia cannot prevent GIs from being co-opted by foreign corporations. Simultaneously, and as indicated in the chapter by Niederle, Wilkinson and Mascarenhas on Brazil, there are some desirable outcomes associated with low institutional strength, whereby the flexibility of a weaker institutional structure allows for the adaptation of GI to match different local realities. Yet the Brazilian case also indicates that institutional flexibility comes at the price of institutional instability. This arises because relatively independent local administrative systems hamper adequate levels of communication and governance. In turn, all of the above is affected by the notion of quality of action. As indicated by Nizam in her discussion of case of GI in Turkey, institutional actions may be branded as effective. Yet effectiveness, as in the case of top-down administrative decision-making processes, does not necessarily translate into the just distribution of added value among stakeholders. In summary, the contradictory outcomes arising from varying institutional strengths and orientations make it difficult to ensure consistently positive implementations of GI.

The issue of protection from the unwanted consequences of the functioning of the market

As illustrated in the introduction chapter, debates on GI stress that a significant part of its emancipatory power rests on the containment of the undesirable consequences of the functioning of the market system. The empirical chapters included in this volume speak directly to this issue and provide insight into the potential of GI to resolve the contemporary agrarian question and the environmental sustainability question. The findings from each case are complex and contradictory, as evidence that GI improves the socioeconomic conditions of stakeholders is often accompanied by equally strong signals that GI contributes to or fails to counteract some of the negative outcomes of the functioning of the market. Accordingly, the third conclusion of the book is that *GI does not necessarily represent a safeguard against the working of the market.* It, however, represents a reaffirmation of the view that markets should be controlled. As discussed in various chapters, and particularly in the chapter by Blancaneaux in his analysis of French wine, the origins of GI rest on the desire to stabilize markets and restore confidence in the fairness of the system. Arguably, the entire GI project is based on the belief that markets should be fair and that producers should get a just reward for their labor. Also implicit in the rationale behind the establishment of GI is the acceptance of the tenets of Keynesian economics and the rejection of the neoliberal argument in favor of deregulated markets. As indicated by Bonanno, the shaping of the market in terms of political objectives (Keynesianism) prioritizes politically determined socioeconomic objectives over the neoliberal preference for self-regulating markets.

The issue of the effectiveness of GI as an instrument to regulate markets is at the core of the analyses presented in the volume. For all contributions, the concept of emancipation rests largely on the achievement of just rewards for producers, the prosperity of their communities, the availability of quality food for consumers and the sustainable use of natural resources. Achieving these objectives, however, is hampered by the characteristics of the current agri-food system with its global competition and dominance of "food from nowhere." As indicated by Feuer, GI has emerged as a promise to local producers in newcomer countries of a requalification of their products and a mechanism for escaping the global mainstream processes of standardization and commodification of agri-food. Moreover, and as illustrated by Appavoo and Korzun, arguments in favor of GI, including economic benefits to producers, the viability of rural economies and the transparency about products and their production, are attractive features to many struggling minorities, not only artisanal producers. In effect, the analyses presented in the book trace how GI is optimistically characterized as a way to allow producers to attain functional and just access to world markets and to subscribe to a system that guarantees traditional and quality forms of production (e.g., Feuer; Nizam; Piatti and Dwiartama; Sekine; Shtaltovna and Feuer; Weiler – all this volume). The negative effects accompanying GI's positive contribution to the control of markets are in turn characterized by GI promoters as manageable or at least fairly compensated for.

However, the extent to which negative effects of the market can indeed be mitigated is difficult to predict. For instance, Blancaneaux underscores in his review of the highly mature case of French wine that market instability and the development of mass-produced, poorly differentiated GI products could not be countered. Similarly, the dominant market position occupied by large corporations or external actors is difficult to oppose. The cases presented by Feuer (Cambodia) and by Piatti and Dwiartama (Indonesia) illustrate that GI remains an inconsistent tool for developing countries to leverage against foreign investors seeking to appropriate locally generated value. These cases also highlight that GI, when applied to former colonial value chains such as coffee and pepper, may even engender forms of neocolonialism. Piatti and Dwiartama, along with Feuer, demonstrate how limited social change, the inadequate distribution of socioeconomic rewards and persistent unequal exchange represent sufficient reasons to question established narratives that understand GI as an instrument of emancipation. They indicate that this is particularly the case in developing countries, which are facing not only powerful domestic actors but also entrenched international presences. A variation of this problem is uncovered by analyses from middle-income countries, such as those of Renard and Domínguez Arista (Mexico) and of Niederle, Wilkinson and Mascarenhas (Brazil), which show the discriminatory effects of GI in favor of large politically connected producers. This problem extends to advanced countries, as represented by Appavoo and Korzun (Canada), Sekine (Japan) and Weiler (Canada and the United States), where the authors underscore that the implementation of GI does little to unseat highly established producers/corporations from positions of power in the food value chain. These authors call for broader reforms, as current GI programs are inadequate to reach declared objectives. Particularly relevant is the limited potential of GI to provide a solution to issues of labor relations and labor remunerations. Sekine demonstrates that GI offers, at best, only partial solutions to the improvement of the working conditions of producers (tea farmers) in general and wage workers (tea pickers) in particular. Echoing Sekine, Weiler shows that the viability of GIs, as represented in the new artisanal food sector, can even depend on labor exploitation.

Aggregation, mobilization and conflict

Discussions on the emancipatory dimension of GI (see introduction chapter) underscore its power to bring about social and political aggregation among actors involved in the production and consumption of traditional or local food. In these discussions, social and political aggregation, or collective action, is considered an important tool for the empowerment of stakeholders, and this rests on at least two aspects of GI. The first refers to traditional know-how that emerges as a binding factor not only among producers but also between producers and consumers as traditional ways to produce food are sought and deemed superior to mainstream production and products. The second refers to the importance of locality that materializes through the recognition that specific places offer special resources that cannot be available and/or replicated elsewhere. These two aspects generate emancipatory outcomes because they create movements that aim at favoring local actors, changing undesirable socioeconomic relations, safeguarding traditional culture and more. Renard and Domínguez Arista; Sekine; Hegnes and Amilien; Niederle, Wilkinson and Mascarenhas; Shtaltovna and Feuer and others all show the importance that the *process* of creating GIs has in fostering the formation of movements for the advancement of the interests of local producers and sympathetic consumers, even if the establishment of a GI is not the final outcome.

GI is also recognized as a shared point of reference that affects the development of consciousness and feelings of membership not only among producers but also for consumers as they feel empowered to be able to choose desired food products and participate in alternative, perhaps shorter, value chains. As chapters in this book stress, enhanced communication among stakeholders strengthens community formation. Hegnes and Amilien, as well as Appavoo and Korzun, indicate that the establishment of novel forms of communication, including the consummation of new vocabulary about food-cultural knowledge, allows for a better understanding of the important links among people, places and food. They add that this process can foster wider engagement. Following the conclusions of other chapters, they state that enhanced communication improves the interaction between producers and consumers to the benefit of both groups, or even enhances societal relations.

Yet GI-generated social aggregation is not free of problems; instead, it is frequently accompanied by issues such as distrust and even conflict. As indicated by Renard and Domínguez Arista; Nizam; Piatti and Dwiartama; and Niederle, Wilkinson and Mascarenhas, GI implementation often requires the divisive creation of artificial boundaries and the consequent difficult task of grouping together heterogeneous actors through the creation of cultural, political and economic common denominators. As highlighted by Feuer; Nizam; Renard and Domínguez Arista; Piatti and Dwiartama; and Sekine, attempts to create new groups and group membership and identity often clash with local stakeholders' existing interests. Particularly problematic is the translation of abstract ideas about a product into codes of practice in which specific standards for GI products are established. While agreement may be reached on the proposal to create a GI product, the establishment of the specific characteristics of this product emerges as a contested terrain since benefits may not be equally distributed among producers. The case of mezcal, illustrated by Renard and Domínguez Arista, further stresses that problems do not simply end with exclusion but instead can escalate to the more serious forms of overt discrimination. These analyses allow us to reach our fourth conclusion of the book:

while GI represents an aggregating force, it also produces centrifugal forces that engender division and conflict.

The protection of the environment and sustainable use of natural resources

Protecting the environment and sustainably using natural resources are among the outcomes often mentioned as desirable byproducts of GI. The rationale behind this claim rests on the assumption that traditional production is not only endowed with a light ecological footprint but also associated with environmentally friendly sensibilities. This rationale is reaffirmed in analyses presented in the book, such as the chapter by Appavoo and Korzun, which offers GI as a solution to combined social and environmental problems affecting minorities in Canada. Simultaneously, however, other analyses reach opposite conclusions, once more underscoring the contradictory nature of GI. One of these instances is the case of the GI for matcha, a powdered green tea analyzed by Sekine. Specifically probing the ecological impact of the establishment of this GI, she stresses not only that ecological requirements are not inscribed in the code of practice but also that GI does not intentionally contribute to biodiversity or aim to reduce the use of agro-chemicals. A more serious example, from Renard and Domínguez Arista, suggests that the popularity engendered by the GIs for mezcal and tequila has directly led to unsustainable wild harvest of agave in Mexico. In essence, the fifth conclusion of the book is *that as far as the protection of the environment is concerned, GI remains an unsystematic and contradictory measure.* Its potential to encourage more sustainable use of natural resources and safeguard of the environment coexist with instances in which alternative approaches to the use of natural resources and the environment are neither contemplated nor practiced.

The existential question

In various forms, all the chapters in this volume indicate that GI is employed as an instrument to support the livelihood of local producers and realize their desire to remain in farming and food production. In this representation, these people want to maintain their ways of life despite highly competitive markets, a corporate dominated agri-food system and adverse socioeconomic phenomena such as globalization and neoliberalism. GI, in other words, emerges as part of the process of affirming identities (i.e., artisanal food producer, small family farmer, Indigenous producer and conscientious consumer) and the political struggle to preserve ways of life that are not only threatened but also on the verge of disappearing. Appavoo and Korzun; Nizam; and Shtaltovna and Feuer offer GI as imperfect forms of protection of indigenous ways, while Hegnes and Amilien view GI as a tool that empowers traditional farming communities. These chapters represent policy contexts in which the GI framework has nominally been erected to encourage local producers to exercise self-determination, be protected market players and maintain the freedom to choose a path of existence that transcends economic necessity and offers alternatives to the constraints established by external forces.

Presented in these terms and despite many of the limits discussed in this book, GI represents for many an emancipatory mechanism that opposes the logic of the free market and its standpoint that only profitable economic activities deserve moral and social support. In this context, GI symbolizes a rejection of the processes of alienation that reduce all labor activities to merely economic events. Alienated labor is deprived of the ability to interpret work in noneconomic ways and select priorities that transcend efficiency, profitability and calculability that are functionally associated with making profit. Unalienated labor, conversely, is free to imagine and pursue different priorities that include the safeguarding of history, tradition and culture; the valorization of human activities and desires; and the fulfillment of people's aspirations. Often, GI products are more expensive and require more production time than their "mainstream" counterparts. However, the GI proposal aims at employing logics that are different from the instrumental rationality of seeking profit. Under these different logics, stakeholders are emancipated to pursue what they *feel* is "right" rather than what *is* right because it is economically profitable.

Simultaneously, the chapters of the book underscore how this emancipatory aspect of GI contains at least two fundamentally contradictory dimensions. The first refers to the unavoidable commodification entailed by GI, which provides a constant pressure for homogenization that becomes, as indicated in some of the cases analyzed in the volume, a dominating effect. The chapters illustrate that stakeholders come to see GI products as commodities that, as such, should be sold at the highest possible price and quantity and to the greatest number of consumers. Additionally, the ability of GI to address local issues is conceptualized in terms of its ability to generate additional revenue. Placed in these terms, GI remains framed in the commodified dimension of dominant social relations and, consequently, it loses a significant amount of its alternative power. The second dimension refers to instances in which stakeholders' self-determination is conditioned by the requirements to remain faithful to tradition and traditional ways of life. In this case, the creation of GI is understood in terms that forfeit change and the potential to overcome undesirable conditions associated with the support of tradition. While the negotiated character of the social creation of tradition permits some leeway here, the conservative dimension of the rootedness to tradition is also explicit. Thus, it tends not only to constrain the freedom of stakeholders to embrace new forms of production but also to readily adopt new social arrangements that may improve their overall well-being.

GI and comprehensive change

The research question of this volume probes whether GI represents a progressive alternative to the socioeconomic conditions that characterize the contemporary agri-food system. As globalization and neoliberalism shape social and industrial relations, centrifugal forces have altered established equilibria and created new and arguably more severe challenges for historical agri-food systems. In this context, the relevance of GI in influencing the conditions of agri-food can be explored through its capacity to resolve two historical questions: the agrarian question and the environmental sustainability question.

The agrarian question

The agrarian question refers to the conditions of people who work in farming and food production. Formulated in the late 1800s, Karl Kautsky sought in his classic book *Die Agrafrage* (the Agrarian Question) (1988 [1899]) to address the issue of the well-being of farmers and peasants given the inadequacies of the then-dominant theories of the evolution of agriculture. At one pole of this debate stood the classic laissez-faire theory in which the coexistence of family farms and large industrial farms were a constant and indeed a feature of the free functioning of the market. According to this theory, the free market allows a sufficient range of new entries into farming such that a natural coexistence of family and large farms would reproduce a multifaceted farming sector. Simultaneously, open competition would encourage processes of internal reconfiguration, allowing small and medium-size operations to expand and transform into large industrial farms featuring mechanization and the use of industrial inputs and wage labor. As illustrated by classic laissez-faire theorists, the pursuit of self-interest on the part of farmers would guarantee the existence not only of efficient forms of production but also of the sustainable use of natural resources (Smith 1981 [1776]). At the opposite pole of the debate, the classic Marxist-Leninist view of farming proposed the theory of the proletarianization of family farmers and the concomitant emergence of green factories whereby urban manufacturing's capital concentration and centralization would be reproduced in the farming sector (Lenin 1961 [1908]). In this view, the disappearance of family producers is inevitable, as they would be transformed into wage laborers. Accordingly, policies designed to improve the living conditions of farmers were to be discouraged because they simply delayed their unavoidable proletarianization.

For Kautsky, and likeminded social democrats, both these positions were incorrect, as they did not contextualize the existence of family farmers and peasants in the overall evolution of capitalism and its necessary equilibria. Kautsky contended that family farming performs several functions in the capitalist economy. One of them is that of a reservoir of labor for large farms and for the manufacturing sector. As additional labor is needed in manufacturing and agriculture, family farm and peasant labor are released to satisfy this demand. The opposite occurs in periods of crisis when surplus labor is retained on the farm, thus contributing to social and economic stability. Following this analysis, Kautsky concluded that because of the functions that they perform, family farms would continue to persist despite economic hardship and capital concentration.

Additional arguments about the relevance of family farming were proposed by theorists sympathetic to the position of Kautsky. In this vein, some argued that family farmers and peasants' paramount desire to own land allows them to endure high levels of socioeconomic hardship. This is the case of the theory proposed by Max Weber. He contended that peasant farming is characterized by high levels of labor self-exploitation. This is a condition, he continued, that allows peasants to pursue the ownership of farmland. Rather than following an economic rationality, peasants obey a different logic that privileges their desire to become owners of their farms (Weber 1958). Alexander Chayanov's *Theory of Peasant Economy* (1986) offers another instance of the different logic that typifies the behavior of small farm holders. For Chayanov, peasants are primarily interested in subsistence production. Accordingly, the quantity of per capita labor performed on the farm is a function of the number of family members and their ages. The younger and more numerous the children in the family are, the more labor is necessary to support them. Once these children become adults and can work on the farm, peasants diminish the per capita quantity of labor performed. In essence, for Chayanov, peasants are not interested in generating surplus value and, therefore, it would be difficult for them to follow the logic of capitalist accumulation.

Throughout the 20th century and the first decades of the next century, this classic debate evolved significantly, primarily due to the drastic reduction of the size of the agricultural labor force and the increased efficiency of agricultural production. Increasingly sophisticated means to control output and modulate agri-food commodity prices, coupled with increasingly assertive world trade, caused what had been deemed economic crises to become issues of national interest. In this context, the agrarian question has been transformed from an issue of universal relevance to a problem that remains significant in some regions and national contexts. It is also an issue that is no longer necessarily economic. Its social dimension has become quite relevant, as exemplified by instances such as the aging of the farming population, the concomitant lack of interest in farming by younger generations, the understanding of farmers as stewards of the land/environment and the rising interest in local, GI and artisanal food.

As the complexity of these phenomena makes sweeping generalizations problematic, proposals that could address the agrarian question clearly cannot be limited to the sectorial space of farming. In other words, throughout its history, the agrarian question has never been about agriculture per se. But its resolution has involved the reorganization of entire economy and society (Ray 1998). This dimension has not changed, and therefore, it requires that solutions to the agrarian question be framed in proposals that involve changes for the entire society. More specifically, issues concerning labor relations, the use of natural resources, the production of safe and adequate food and its distribution and demand cannot be simply addressed by generating measures that are exclusively confined to the agri-food sector. While proposals focusing on the sectorial dimension of agri-food will more likely constitute a significant portion of the intellectual and political production, a meaningful restructuring of agri-food should transcend the sectorial sphere (Mooney 2004). Following this point, GI has not been able to be, by itself, an instrument that would engender significant changes in agri-food. Given the nature of current labor and production relations, the expansion of GI and its improvement only creates relatively marginal alterations in the functioning of capitalist markets and agri-food production. Although this assessment perhaps asks GI to achieve an unfairly expansive scope of action, we state it here as a counterpoint to appraisals of GI that do believe in its capacity for resolving and counteracting structural constraints. Similar considerations are pertinent for the environmental sustainability question.

The environmental sustainability question

While the challenge of safeguarding the environment also transcends the sphere of agri-food, the clarifying question to pose for agri-food production concerns the extent to which production practices (regardless of how historical they are) allow for an objectively sustainable use of natural resources (Belletti, Marescotti and Touzard 2017). As illustrated in the chapters of this volume, the realization of environmental sustainability question requires more than relative improvements within the scope of the agri-food sector, let alone the scope of GI policies. These findings are not surprising when seen against classic and contemporary debates on how best to encourage a safeguarding of the environment.

Classic economic theory proposes the market equilibrium solution to the environmental question whereby the increased use of natural resources is controlled by price changes. This is also the solution proposed by classic neoliberals. Criticizing environmentalists' arguments about the excessive use of natural resources by mainstream industries, Friedrich von Hayek (2011 [1960]) contended that demand-generated price increases stimulate the search for further cost reductions and efficient production techniques. In the case of contemporary environmental limits, neoliberals argue that the environmental issue is effectively addressed through market mechanisms (Friedman 1970; Jordan, Wurzel and Zito 2004). This phenomenon is substantiated, while also revised, in some fashion in Henderson's (1999) analysis of capitalist agricultural development in California. Henderson documents how apparent environmental limits were consistently overcome through major capital investments and concurrent technological upgrading, such that California could transform hitherto unproductive lands into a global agricultural commodity producer. This was achieved by the fact that California, as a pioneer agricultural zone, lacked the usual "drag" of historical agri-food systems and could, as proposed by Hayek, achieve increasing efficiency. Seen inversely by critics, the capitalist development of California agriculture entailed internalizing forms of environmental and labor exploitation that would soon (and increasingly thereafter) be re-evaluated as problematic externalities.

It is therefore important to point out that the neoliberal view departs from the classic laissez-faire theory proposed by Adam Smith and 19th century philosophers, such as Emile Durkheim, who were seeking an answer to the environmental question. Here, two general principles are particularly relevant for the establishment of desirable economic growth and a balanced society. The first refers to the importance of the human sentiment of sympathy. Sympathy, Smith contended, must accompany competition so that the pursuit of self-interest goes always together with the respect of others and their interests (Smith 2010 [1759]:81). This means that the framework within which self-interest should be sought is defined by the collective understanding of what is considered good by the entire society, an idea echoed one century later by Durkheim (1993 [1887]). Stressing the importance of the role of the State in defining national priorities, Smith saw the sphere of moral obligations, such as environmental conditions, framing economic goals, while Durkheim (1984 [1933] believed a sustainable equilibrium ought to be reached through the conscious implementation of appropriate political measures (social facts) (Durkheim 1984 [1933]). The second principle refers to the need for a "just" socio-economic system. The prospering of the economy, Smith believed, cannot be achieved unless justice informs the organization of society. In this context, economic activities must be subordinate to the continuous existence of a just society. This type of society is one formed of moral people who place the overall interests of society above their own and avoid situations in which their actions create harm to all aspects of society, including the environment.

For Max Weber (2002 [1905]), the evolution of capitalism is based on the continuous rationalization of social relations. Rationalization signifies the application of formal rationality – or the instrumental use of all resources – to achieve capital accumulation. While rationalization advances capitalism, it also creates significant problems as it eclipses the adoption of other forms of rationality, such as substantive rationality or acting based on values. Accordingly, the rational exploitation of the environment, although fundamental for the expansion of capitalism, remains a significant liability for the future of society and thus requires alternatives. Weber further stressed the importance of thinking about the future of society in terms of the application of substantive rationality. This form of rationality was practiced in the past but has progressively disappeared in contemporary society. In this context, the call for protecting the environment assumes a fundamental dimension in the construction of a better and sustainable future.

In the case of Marx and the Marxist tradition, the connection between the resolution of the environmental sustainability question and change that involves the entire society is made overtly clear. The major tenet of the Marxist position is the unsustainability of capitalism as it engenders not only the uncorrectable exploitation of labor but also of the environment. Illustrated by Marx in his discussion on "Large Scale Industry and Agriculture" in *Capital Volume I* (1977, 636–639), the contradiction between the evolution of capitalism and the safeguarding of the environment rests on the very implantation of capitalist social relations and the expansion of industrialization and urbanization that it engendered. Popularized by contemporary works on Marx's "Metabolic Rift" and the "Second Contradiction of Capitalism" (Bellamy Foster 1999; Moore 2011; O'Connor 1988), this theory contends that pre-capitalist societies were based on a stable and sustainable metabolic interaction between cities and the surrounding rural (farming) areas whereby the natural resources (i.e., nutrients) necessary for the production of food for the cities were recycled to the countryside as waste. While preventing advanced urbanization, the time reduced technological development placed balancing limits on the exploitation of natural resources. With the growth of capitalism, these sustainable conditions were forever altered. The expansion of industrialization and the concentration of population in large urban areas mandated the extraction of an unprecedented quantity of raw material and the production of an equally large quantity of food. As the growth of cities required increasingly large quantities of nutrients, balancing amounts were not returned to the countryside, causing an unequal exchange between urban and rural areas. As a result, urban pollution developed in tandem with the depletion of natural resources. In essence, for Marx and the Marxist tradition, the environmental crisis is intrinsically linked to the unsustainability of capitalism.

The extent to which GI, which is an instrument designed to latch onto capitalism while mitigating unwanted economic and social consequences, can engender environmental values is undermined by its built-in respect for the preservation of traditional or historical practices and livelihoods, regardless of their objective environmental performance. This historical review of the role of the environment in capitalist development demonstrates unequivocally that sustainability is unlikely to be achieved as a matter of course in the development of agri-food. As a consequence, the environmental performance of various GI products is not systematically assured but is instead idiosyncratic and dependent on broader conditions in society. This echoes the judgment reached concerning the agrarian question, namely that GI cannot play a comprehensive emancipatory role in the resolution of the environmental sustainability question. The requirement of comprehensive change essential to the resolution of these two questions simply transcends the sectorial dimension of GI. Additionally, GI remains a market-based instrument. Therefore, arguments that identify the origins of agri-food problems in the inadequacy of the capitalist system are likely to find GI an ineffective instrument to address the structural conditions of a market-based society, as highlighted by the Marxian and Weberian traditions. Simultaneously, however, while the limited emancipatory dimension of GI is a contention shared by the analyses in the volume, the chapters document that the sectorial dimension of GI offers numerous "partial" yet relevant outcomes.

The emancipatory power of GI

The overarching conclusion derived from this volume is that the existing *emancipatory power of GI, which represents a beacon of hope to many agri-food stakeholders, is contradictory because it is diminished by the very factors that enable it.* This general conclusion can be applied to the overall experience of GI illustrated in this book. GI is predominantly a market mechanism that is employed to combat the negative outcomes of the functioning of the global market. Its deployment, however, occurs in a context in which the mechanisms that define market functioning are not substantively altered. In essence, GI does not have the scope of action to modify the overall functioning of the market. And yet, it is an instrument that is over-optimistically charged by promoters with achieving a number of objectives that are alternatives to the currently dominant character and conditions of the agri-food system. In this case, however, the contradictory aspect of GI arises from its inability to ultimately promote significant alternatives to market relations or escape structural constraints imposed from outside the agri-food sector. This implies that the conditions that engender fundamental market trends such as capital concentration and centralization, marginalization of smaller producers and the exploitation of natural resources are not comprehensively changed. To some extent, though, the emancipatory dimension of GI appears undeniable, because it promotes processes that do mitigate certain negative consequences of the dominant agri-food system. Therefore, and despite its limits, GI can be considered a "least-bad" tool that should be employed and constantly refined to achieve a more economically just, socially acceptable and environmentally friendly agri-food system. The evidence provided in the chapters of this volume provide insights about how to proactively refine GI policies, but it also reminds us to remain sanguine about the potential for truly emancipatory outcomes. The conditions under which an emancipatory role could be played is arguably a topic that deserves further investigation.

Bibliography

Bellamy Foster, John. 1999. "Marx's Theory of Metabolic Rift: Classical Foundations for Environmental Sociology." *American Journal of Sociology* 105 (2): 366–405.

Belletti, Giovanni, Andrea Marescotti, and Jean-Marc Touzard. 2017. "Geographical Indications, Public Goods, and Sustainable Development: The Roles of Actors' Strategies and Public Policies." *World Development* 98: 45–57.

Chayanov, Alexander. 1986. *The Theory of Peasant Economy*. Madison: University of Wisconsin Press.

Durkheim, Émile. 1993 [1887]. *Ethics and the Sociology of Morals*. Buffalo: Prometheus Books.

———. 1984 [1933]. *The Division of Labor in Society*. New York: The Free Press.

Friedman, Milton. 1970. "The Social Responsibility of Business Is to Increase Its Profits." *New York Times Magazine*, September 13, pp. 32–33, 122–26.

Hayek, Friedrick. 2011 [1960]. *The Constitution of Liberty*. Chicago: University of Chicago Press.

Henderson, George L. 1999. *California and the Fictions of Capital*. New York: Oxford University Press.

Jordan, Andrew, Rudiger K. W. Wurzel, and Anthony R. Zito, eds. 2004. *New Instruments of Environmental Governance? National Experiences and Prospects*. London: Routledge.

Kautsky, Karl. 1988 [1899]. *The Agrarian Question*. London. Pluto Press.

Lenin, Vladimir Ilich. 1961 [1908]. *Theory of the Agrarian Question*. London: International Publishers.

Marx, Karl. 1977. *Capital Volume I*. New York: Vintage Books.

Mooney, Patrick H. 2004. "Democratizing Rural Economy: Institutional Friction, Sustainable Struggle and the Cooperative Movement." *Rural Sociology* 69 (1): 76–98.

Moore, Jason W. 2011. "Transcending the Metabolic Rift: Towards a Theory of Crises in the Capitalist World-Ecology." *Journal of Peasant Studies* 38 (1): 1–46.

O'Connor, James. 1988. "Capitalism, Nature, Socialism a Theoretical Introduction." *Capitalism, Nature and Socialism* 1 (1): 11–38.

Ray, Christopher. 1998. "Culture, Intellectual Property and Territorial Rural Development." *Sociologia Ruralis* 38 (1): 3–20.

Smith, Adam. 2010 [1759]. *The Theory of Moral Sentiments*. New York: Penguin.

———. 1981 [1776]. *An Inquiry into the Nature and Causes of the Wealth of Nations*. Indianapolis, IN: Liberty Classics.

Weber, Max. 2002 [1905]. *The Protestant Ethic and the Spirit of Capitalism*. New York: Penguin.

———. 1958. "Capitalism and Rural Society in Germany." In *Form Max Weber: Essays in Sociology, Capitalism and Rural Society in Germany*, edited by Hans H. Gerth and C. Wright Mills, 363–85. New York: Oxford University Press.

Index

Note: Page numbers in *italics* indicate figures, in **bold** indicate tables.

For Product Safety Concerns and Information please contact our EU
representative GPSR@taylorandfrancis.com
Taylor & Francis Verlag GmbH, Kaufingerstraße 24, 80331 München, Germany

www.ingramcontent.com/pod-product-compliance
Ingram Content Group UK Ltd.
Pitfield, Milton Keynes, MK11 3LW, UK
UKHW021722240926
475702UK00029BA/1796

* 9 7 8 0 3 6 7 7 2 7 1 1 6 *